The Romans

Roy Burrell   Illustrated by Peter Connolly

Oxford University Press

Oxford University Press, Great Clarendon Street, Oxford OX2 6DP

Oxford New York
Athens Auckland Bangkok Bogota Bombay
Buenos Aires Calcutta Cape Town Dar es Salaam Delhi
Florence Hong Kong Istanbul Karachi
Kuala Lumpur Madras Madrid Melbourne
Mexico City Nairobi Paris Singapore
Taipei Tokyo Toronto

and associated companies in
Berlin Ibadan

Oxford is a trade mark of Oxford University Press

© Text: Roy Burrell 1991
© Illustrations: Peter Connolly 1991

First published 1991
Reprinted 1997

Library of Congress Catalog Card Number 90-53259

A CIP catalogue record for this book is
available from the British Library

ISBN 0 19 917102 5 pbk

Typeset by MS Filmsetting Limited, Frome, Somerset
Printed in Hong Kong

Acknowledgements

The illustrations including the cover and map models are by
Peter Connolly. Handwriting is by Elitta Fell.

The publishers would like to thank the following for permission
to reproduce photographs:

Robin Birley/The Vindolanda Trust p. 61; Dorset Natural
History Society, Dorset County Museum, Dorchester, Dorset
p. 91; Robert Estall p. 111; Michael Holford p. 32; The
Hutchison Library p. 110; Ny Calsberg Glyptotek, Copenhagen
p. 47; Rheinisches Landesmuseum Trier p. 60; Scala p. 11,
p. 16, p. 23, pps. 30/31, p. 37, p. 38, p. 45, p. 48, p. 49, p. 55,
p. 72, p. 74; Ronald Sheridan's Photo Library p. 49 bottom,
p. 100; Weidenfeld and Nicolson Ltd./Galleria Borghese p. 83.

CONTENTS

The legend of Aeneas

On this page and the next are two of the stories that Roman parents told their children when they asked, 'Who were the earliest Romans? Where did they come from?'

There were many legends about the characters who fought in the Trojan War. Virgil, a poet of later Roman times, wrote about Aeneas, a warrior hero, and ancestor of the Roman people. He told the story of Aeneas in a long poem called the *Aeneid*.

Virgil begins by describing how the Greeks attacked and captured the city of Troy. Some Trojans, however, managed to escape. Aeneas, his father Anchises, and his son Ascanius, arranged to meet at Mount Ida, a low hill overlooking the city. Creusa, Aeneas's wife, never arrived. Aeneas sadly concluded that she must have become separated from the family in the stampede of refugees, and had probably been killed in the crush of people.

The family made their way to Antandros, where they managed to get aboard a ship that was fleeing from the doomed city. They sailed to Greece and to the island of Crete, but Aeneas wanted to start building a new city of Troy, and he was sure that neither of these was the right place to do so.

At one point they were attacked by harpies, a flock of birds with human heads who stole their food. The leader of the harpies spoke mockingly to them, saying that they wouldn't find the right place for their new city until they had been forced by hunger to eat the tables on which their food was spread.

They tried to sail from there to Italy, but were blown off course by storms and forced ashore in North Africa. Queen Dido, the ruler of the nearby city of Carthage, took pity on the Trojans. She fell in love with Aeneas and tried to persuade him to marry her. But Aeneas felt he had to move on to look for the place to build the new Troy. When he sailed away, Dido killed herself.

Finally, the wanderers came to the end of their voyage at the mouth of the river Tiber in Italy, where they disembarked.

As they were eating, one of them looked at the meal of pieces of meat laid on slices of wheatcake and exclaimed, 'Aeneas! The prophecy of the harpy queen

Aeneas escapes from Troy with his father and son

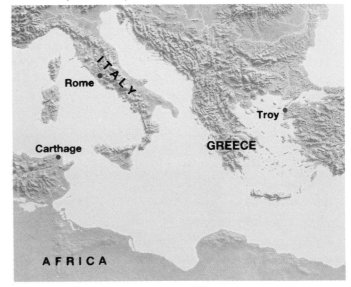

The death of Dido

has come true: are we not eating the tables on which our food is spread?'

They found out that the country was called Latium, and they set out to explore it. Before long, they met Latinus, the king, and Aeneas fell in love with his daughter, Lavinia. Aeneas told the king that he wanted to rebuild Troy, and asked for Lavinia's hand in marriage.

Turnus, the ruler of a nearby land, also wanted to marry Lavinia. When Aeneas wouldn't give way, he challenged him to fight for her. Aeneas had the worst of the duel to start with, but eventually he managed to break Turnus's sword, and the fight was over.

Aeneas built his new town, and called it Lavinium. He and his wife reigned there for many years. After the death of Aeneas, his son Ascanius became king. Ascanius decided to build a new city of his own, called Alba Longa. He left Lavinium and made Alba Longa his capital.

Virgil's account of these adventures ends with the death of Turnus, but other Roman writers tell how the descendants of Ascanius reigned in Alba Longa for over over four hundred years.

Aeneas lands at the mouth of the Tiber

The legend of Romulus and Remus

NUMITOR WAS THE DESCENDANT OF AENEAS. HE RULED IN ALBA LONGA SOME FOUR CENTURIES AFTER THE DEATH OF HIS FAMOUS ANCESTOR.

AMULIUS WAS JEALOUS OF HIS BROTHER, KING NUMITOR, AND RESOLVED TO TAKE HIS THRONE BY FORCE.

AMULIUS DROVE OUT NUMITOR AND REIGNED AT ALBA LONGA IN HIS STEAD.

RHEA SILVIA WAS NUMITOR'S ONLY CHILD, BUT AS A GIRL SHE SEEMED TO PRESENT NO DANGER TO HER WICKED UNCLE.
TO PREVENT HER MARRYING AND PRODUCING A SON WHO MIGHT CLAIM THE THRONE, AMULIUS FORCED HER TO BECOME A PRIESTESS OF A TYPE WHO WERE FORBIDDEN TO MARRY.

IN SPITE OF THESE PRECAUTIONS, MARS, THE ROMAN WAR GOD, MARRIED HER SECRETLY AND SHE HAD TWIN BOYS NAMED ROMULUS AND REMUS.

AMULIUS ORDERED A SERVANT TO DROWN THE TWINS IN THE RIVER, AND RHEA SYLVIA TO BE PUT IN PRISON.

THE SERVANT TOOK PITY ON THE BOYS AND FLOATED THEM DOWN THE RIVER TIBER IN A LITTLE WOODEN CRADLE.

THE CRADLE CAME TO REST AT A PLACE NEAR SEVEN HILLS. A SHE-WOLF WHOSE CUBS HAD BEEN KILLED FOUND THE TWINS. HER BODY WAS HEAVY WITH MILK, SO SHE FED THEM.

7

FAUSTULUS, ONE OF THE ROYAL SHEPHERDS, FOUND THE CRADLE AND TOOK THE BABIES HOME. HE AND HIS WIFE SECRETLY BROUGHT THE BOYS UP AS THEIR OWN CHILDREN.

ROMULUS AND REMUS GREW UP TO BE STRONG AND ATHLETIC. THEY BECAME THE LEADERS OF THE LOCAL YOUTHS.

WHEN THEY BECAME MEN, THEY FOUND OUT WHAT THEIR GREAT-UNCLE HAD DONE AND THEY ATTACKED HIS CITY. AMULIUS WAS KILLED IN THE FIGHTING. NUMITOR WAS RESTORED TO HIS THRONE, AND THEIR MOTHER, RHEA SILVIA, RELEASED FROM PRISON.

THE YOUNG MEN DECIDED TO BUILD A CITY ON ONE OF THE SEVEN HILLS NEAR WHERE THE WOLF HAD FOUND THEM. THEY COULDN'T AGREE WHICH HILL TO BUILD ON, SO THEY SAT AND STARED AT THE SKY. WHOEVER SAW A VULTURE SHOULD HAVE HIS CHOICE.

ROMULUS STAMPED OFF AFTER THE ARGUMENT, BORROWED A PLOUGH AND OX TEAM, AND CUT A FURROW ROUND THE PALATINE HILL.

REMUS SAW SIX VULTURES BUT ROMULUS CLAIMED TO HAVE SEEN TWELVE. REMUS SAID THAT HE HAD SEEN THE FIRST BIRD BUT ROMULUS DECLARED THAT THE CHOICE WAS HIS, SINCE HE HAD SPOTTED MORE BIRDS.

'THIS IS THE FRONTIER', ROMULUS TOLD HIS BROTHER. 'TO CROSS IT WILL MEAN DEATH.' 'LIKE THIS?' ASKED REMUS CONTEMPTUOUSLY, STEPPING OVER THE FURROW. ROMULUS DREW HIS SWORD FURIOUSLY AND SLEW HIM.

ROMULUS WAS SORRY FOR WHAT HE HAD DONE BUT HIS CITY CONTINUED TO GROW. IT WAS NAMED 'ROME' AFTER ITS FOUNDER.

Section 1 *Origins*

The truth behind the legends

One of the difficulties which faces anyone anxious to discover the truth about early Rome is that most of it may be for ever out of reach. The evidence which might have been unearthed is probably buried beneath the remains of a later, classical Rome. It would take a brave archaeologist to suggest demolishing the Colosseum to see what was underneath it. All the same, some digging has been done and a certain amount of evidence brought to light.

The legend of Romulus and Remus is set by Latin

Warriors beside huts on the Palatine Hill

writers almost eight hundred years before the birth of Jesus Christ – at 753 B.C. This may be fairly near the truth, as the earliest hilltop huts and graves found seem to have been made at about this time.

We know that the huts were small and round because of the post holes left in the soil. In addition, small clay models of houses have been dug up from cemeteries. We believe these were used to contain the ashes of cremations. Beneath the surface of what was later to become the Forum, or market square, were found some even earlier burials.

This kind of arrangement – roughly-built huts on lowish hilltops – appears to have been fairly common on the plains of Latium. No one could have guessed that the settlement on the Palatine Hill near the Tiber would one day become the mightiest city of the ancient world.

The story of Aeneas and his escape from Troy is almost certainly an invention rather than real history. The first dwellers in the earliest Latin towns were not heroic warriors from the Greek world, but simple local shepherds or farmers.

It's even possible that the four centuries that were supposed to separate Aeneas from Romulus and Remus were put into the legends deliberately to account for the time that must have gone by between the Trojan War and the founding of Rome.

Another invention must have been the supposed love of Dido, Queen of Carthage, for Aeneas. In fact, Carthage was only founded about a century before Rome – certainly not at the time of the Trojan War. If Aeneas had left burning Troy and landed in North Africa, he would have found no trace of Carthage nor of Dido.

Latin histories of the earliest city of Rome contain much legendary material and are not very reliable. In any case, the histories are much too late to be of great worth. The first Roman we know of to set down the story of Romulus and Remus was Q. Fabius Pictor and he didn't do so until about 200 B.C. Rome had then been in existence for more than five hundred years.

Some people have said that the word 'Rome' must have come from Romulus's name and therefore his existence is proved. In fact, of course, this is no argument, as no one really knows where the city's name came from. Perhaps the name 'Romulus' was invented to account for the name 'Rome'.

Another mistake in the legends concerns all three of the original Latin cities: Lavinium, Alba Longa and Rome. Archaeology shows that the first two were surely as old as the stories say, but Rome, which should have been four centuries younger, was actually proved to be as ancient as Lavinium and Alba Longa.

Romulus is also credited with the organisation of

Hut urn

local government in Rome – but again, there is no proof. It is said that Romulus chose a hundred 'fathers' to help him rule. They formed the first 'senate', or parliament and their descendants were known as 'patricians', from the Latin word for 'father'. He is then supposed to have divided the people into three main tribes and each tribe into ten smaller units called 'curiae'.

After this, the legend says, he devised a system of recruiting soldiers to defend the city. There were to be 3000 infantry and 300 horsemen, a third of which came from each tribe.

This was most likely to have been no more than a spare time army. The soldiers probably followed their ordinary jobs for much of the time. They were plant or stock farmers, keeping cattle and sheep, and growing wheat, barley, peas and beans. Many could only afford to keep pigs and goats for livestock. The vine and olive were as yet unknown in Italy.

Some farmers also doubled as simple craftsmen, making everyday items in wood, metal or clay for themselves and others. All these were humble occupations.

Later Romans, who knew nothing of their ancestors or the founding of Rome, made up stories about the earliest Romans and their adventures, making them seem more exciting than they actually were. This is how legends sometimes begin.

Section 1 *Origins*

The early kings

The legends say that the early city was ruled by a succession of seven kings, beginning with Romulus. Some modern historians have cast doubt on the details of their lives and even on whether they lived at all. It's unlikely that they were completely made up, even though we don't really know what they looked like. They were:

1 Romulus (Latin) (753–718 B.C.) Founded the city and began its system of local government. He also started Rome on its process of growing larger and taking in more and more land. He built up the population by welcoming anyone who wanted to move to Rome.

2 Numa Pompilius (Sabine) (717–673 B.C.) A year after the death of Romulus the new ruler had still not been elected. In the area of Rome there were two main tribes of people – Romans and Sabines – and each wanted the honour. The problem was solved when both tribes agreed that a Sabine should be king and that the Romans should do the choosing. Numa Pompilius reorganised the state religion, and founded colleges for priests, the latter being known as 'flamines'. The flamines were taught how to 'take the auspices'. This meant reading the future from the flight of birds and later from flashes of lightning. He also devised a new calendar of twelve months, replacing one of ten months. The last four of our 'modern' months come from Latin words for seventh, eighth, ninth and tenth.

3 Tullus Hostilius (Latin) (672–641 B.C.) He was a military-minded king who believed that his subjects would become soft if they weren't engaged in fighting every so often. He was so successful in his wars that he began to think that armies were more important than ordinary people, or even the worship of the gods. He extended Rome's rule within a circle roughly ten or twelve miles in radius. Among other neighbouring settlements, soldiers under his command attacked and destroyed Rome's original mother city, Alba Longa. At last the gods grew angry and slew the impious king by striking his palace with a bolt of lightning.

Soldiers of the Servian army

4 Ancus Martius (Sabine) (639–616 B.C.) He extended Rome's boundaries to the coast and captured Ostia, which was to become Rome's seaport. He was probably also responsible for the first bridge over the Tiber. This was the Pons Sublicus, just downstream from the river island opposite the Palatine Hill. In addition, he is supposed to have captured and fortified the Janiculum Hill on the far side of the Tiber.

5 Lucius Tarquinius Priscus (Etruscan) (616–579 B.C.) He was the first Etruscan king. He was appointed tutor to the sons of Ancus Martius. When the old king lay dying, Priscus sent the sons away so that he could have himself named as the next ruler. As king, he subdued more of Rome's neighbouring tribes. He built the Capitol temple on the Capitoline Hill but is perhaps more famous for draining the marshes between the Palatine and Aventine Hills. The sewer he had made was called the Cloaca Maxima ('the Great Sewer') and still runs into the Tiber. On the reclaimed land he ordered the laying out of the Circus Maximus, Rome's first chariot-racing course.

6 Servius Tullius (Etruscan) (578–535 B.C.) He divided the citizens into five classes from the richest to poorest. All except the very poor had to provide standby soldiers for the army. The richest groups supplied the cavalry and the rest supplied the infantry. There were also to be corps of specialists such as

carpenters, metalworkers, signallers and engineers. He is said to have built an enormous defensive wall taking in all seven hills, running for about seven miles and enclosing about one and a half square miles. However, archaeologists say that surviving stretches of Roman town wall are much later than was once thought.

7 Tarquinius Superbus (Etruscan) (534–509 B.C.) The name 'Superbus' means 'proud' and, indeed, he was so haughty he took no notice of what his people wanted or needed. He behaved so badly that the people not only drove him out but promised themselves solemnly that they would never have a king to reign over them again.

Rome in the 5th century B.C. surrounded by the Servian walls

Capitol

Palatine Hill

River Tiber

Section 1 *Origins*

Sabines and Romans

As we've already seen, the first citizens of Rome had to contend with other inhabitants of the various cities, towns and villages on the plains of Latium. One such tribe was the Sabines, a group which provided Rome with some of its first rulers. The legend telling how they came to co-operate with Romulus and his men is a rather interesting one.

It happened in the first few years after the founding of the city. Romulus had issued a general invitation to anyone who wished to make his home in Rome. The more men it had, the more easily it could be defended. As a result, many men were attracted to the hilltop town: unfortunately, many of them were runaway slaves, petty criminals or even murderers!

Although Rome might have grown fast in the early years, as the newcomers arrived, it couldn't go on growing without a good number of wives and families as well. Lacking these, the men would have aged and then died and without children to follow them, Rome would have died too.

The problem was – where were the wives to come from? The daughters of neighbouring tribesmen were unwilling to marry penniless upstarts like these early Romans. After all, many of them had still risen no higher than cattlemen, shepherds or farm labourers.

Someone had an idea – it may have been Romulus or one of his helpers. It was a very simple idea. The Romans were to organise some games in the form of

Rome and its neighbours

athletic contests. All the nearby tribes were invited.

Of course, the Romans, like the Greeks, only allowed men and boys to watch the contests, which left the womenfolk at home and alone. After the races had started, a secret signal was given and many of the Romans quietly left. They stole away to the Sabine settlements and kidnapped most of the Sabine daughters.

Naturally, the Sabine men soon found out what had happened and were furious. They armed themselves and prepared for battle with the treacherous Romans. Romulus and his men were expecting an attack and they were also equipped with weapons and armour. The two forces met just outside the city and drew up in long lines of men, each soldier facing an enemy.

Before the struggle could begin, however, a strange thing happened. Between the hostile groups ran the Sabine daughters – hundreds of them. When they had filled the space which in modern times would be called 'no man's land', they explained what they were doing.

'We were just daughters a short while ago,' said their leader, 'now we are both wives and daughters. We did not choose our husbands – they chose us. We want this fighting to stop. If it goes ahead, many will be slain. When our fathers are dead, we shall be orphans, but if our husbands die, we shall be widows. We lose either way.'

There was silence while this explanation and plea sank in. Finally it was obvious that the women were right and the battle never took place. At least, that's what the legend says and it does explain how Sabine men were able to become kings in ancient Rome.

Of course, the legend could be quite wrong. Some historians claim that these incidents never happened at all. They point to a long, drawn-out war with no obvious winner as the reason for the alternation of the first kings between Latins and Sabines.

The rape of the Sabine women

Section 1 *Origins*

The Etruscans

No one really knows who the Etruscans were nor where they came from. Some historians have suggested Asia Minor as their homeland and in particular, that part which today we call Turkey.

Their civilisation is more of a mystery than that of the Romans because we can't understand very much of their language. We have thousands of examples of their writing but these are nearly all inscriptions on tombs. As such, there aren't enough words in continuous sentences to make much translation possible. We are, however, pretty sure that Etruscan isn't related to any other European language, as French and Spanish are related to each other, for example.

We do know that they inhabited a part of Italy called Tuscany. This lies to the north and west of the river Tiber and Rome. The Etruscans had their own name for it – 'Etruria'.

Fortunately for us, the tombs can tell us a good deal about everyday life, even if we can't read Etruscan writing. The dead were placed in coffins in stone-cut chambers, the walls of which were often painted in bright colours with scenes from the life of the departed.

In addition to the vivid pictures, there were other things from which we can learn about Etruscan ways. In common with many other ancient peoples, the Etruscans often made their tombs resemble the insides

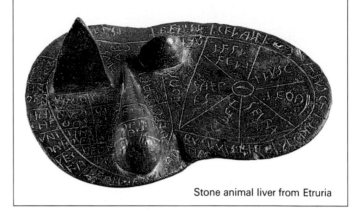

Stone animal liver from Etruria

of their houses. They also buried many items of every-day use to be the possessions of the dead person in the afterlife.

Some of our knowledge of the Etruscans comes from Roman writers. Unluckily, the most complete account of this mysterious people is missing. It was written by the emperor Claudius early in the Christian era but the history, although running to some twenty volumes, has totally disappeared.

Archaeologists can also provide information and from all these sources a dim misty picture begins to take shape. The Etruscans, at least in their early days, were a cultured people with a love of fine things. They wore brightly coloured clothes patterned on those of the Greeks.

In fact, they admired the Greeks a great deal and copied a lot of their culture. Etruscan painted pottery, for example, can often be confused with that from Corinth or Athens. The Romans imitated many Etrus-can things, including the pottery. They adopted a new method of infantry fighting which had come from Greece by way of Etruria.

Many of the ideas which we think of as purely Roman were also borrowed from the Etruscans – for example, the 'triumph' (a procession honouring a victorious general), 'games' (just like the Greek ones), the 'fasces' (axes in bundles of rods as badges of authority) and 'auspices' (telling the future, first of all from birds, then from animal livers and flashes of lightning).

As well as ideas, Rome also imported great quantities

Etruscan writing

16

of Etruscan art work, since the Etruscans were good craftsmen in metalwork, especially in bronze, silver and gold. As well as small items such as rings, bracelets and ornamental jugs, they also made large metal plaques with designs hammered in, which were intended to decorate furniture and chariots. Among the most sought after treasures were bronze mirrors with pictures of people on the back.

The Romans learnt a good deal about practical matters from the Etruscans and even had a century when they were ruled by Etruscan kings.

Towns in Etruria were laid out on a right-angled grid pattern, with houses, squares, temples and market places. They were often sited on hilltops, as was Rome itself.

Etruscan troops from such towns fought against other tribes in Italy. They were beaten by the Greeks in southern Italy and cut off from some of their lands by peoples such as the Aequi and Volsci. The Romans finally subdued them and they disappeared from the scene.

Romans envied the Etruscans their elegance and tried to imitate it. They also disliked and distrusted the men of Tuscany and fought fiercely against them.

The Etruscan town of Marzobotto was laid out on a rectangular grid pattern.

Placing a body in the Campana tomb at Veii near Rome

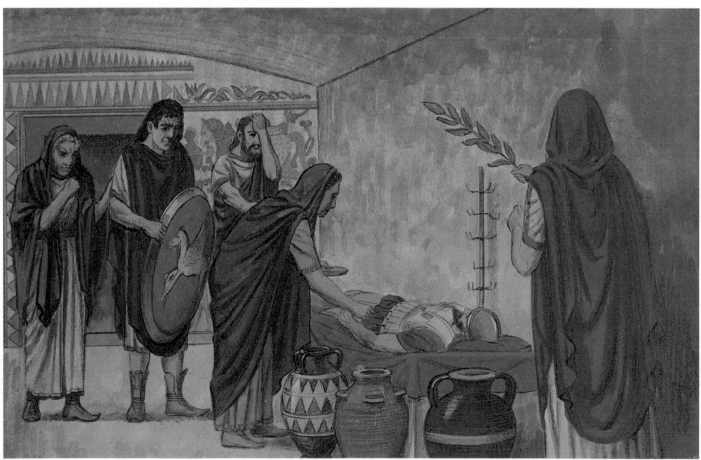

17

Section 1 *Origins*

Horatius and the bridge

In the early nineteenth century, Lord Macaulay wrote a poem celebrating a most heroic feat of arms performed by Horatius, as he defended Rome from its enemies.

After Rome's last king, Tarquin the Proud, had been thrown out of the city, the Etruscans decided to avenge this insult. The poem tells how the Etruscan leader, Lars Porsena, first gathered his army together:

'Lars Porsena of Clusium
By the nine Gods he swore
That the great house of Tarquin
Should suffer wrong no more.
By the nine Gods he swore it,
And named a trysting day,

And bade his messengers ride forth,
East and west and south and north,
To summon his array.'

When the soldiers had mustered, there were 'four-score thousand' infantry and 'ten thousand horse'.

As this huge army neared the city, refugees poured into Rome to shelter behind its walls, and sentinels could see a sky made blood-red with the flames of burning villages. Rome's parliament, the Senate, decided that since the Janiculum Hill, on the far side of the Tiber, had been captured, the only bridge over the river must be destroyed. Before this could be done the guards

Lars Porsena orders the attack on Rome

Horatius holds the bridge

shouted that the Etruscan army was already in sight and that it was too late to smash the bridge.

Then Horatius volunteered to go to the far side of the river with two companions and hold up the Etruscans while the bridge was being chopped down. As he says in the poem:

'To every man upon this earth
Death cometh soon or late.
And how can man die better
Than facing fearful odds
For the ashes of his fathers,
And the temples of his Gods.'

The two friends Horatius took with him were Spurius Lartius and Herminius. They prepared for action whilst the working party got busy knocking down the bridge.

'Now while the Three were tightening
Their harness on their backs
The Consul was the foremost man
To take in hand an axe:
And Fathers mixed with Commons
Seized hatchet, bar and crow,
And smote upon the planks above
And loosed the props below.'

All the Etruscan warriors were beaten in turn by the trio:

'Stout Lartius hurled down Aunus
Into the stream beneath:
Herminius struck at Seius
And clove him to the teeth:
At Picus brave Horatius
Darted one fiery thrust;
And the proud Umbrian's gilded arms
Clashed in the bloody dust.'

Spurius Lartius and Herminius darted back just before the bridge fell, thinking Horatius was with them. When they realised that he wasn't, they tried to return but alas! the bridge had fallen and Horatius was alone. He prayed to the gods and then, in full armour, he dived into the Tiber. Some of the enemy cursed him and hoped that he would drown but Lars Porsena rebuked them and said that such a brave man deserved to survive.

Finally Horatius reached the city bank and staggered ashore, bleeding but alive, and Rome was saved. His fame, the poet says, will resound down the ages:

'And still his name sounds stirring
Unto the men of Rome,
As the trumpet-blast that cries to them
To charge the Volscian home;
And wives still pray to Juno
For boys with hearts as bold
As his who kept the bridge so well
In the brave days of old.'

All very stirring stuff about how heroic the young men of Rome were and how glorious Rome's history. However, some historians maintain that the whole story is false: that Lars Porsena actually conquered Rome and ruled there for some years.

It's certainly difficult to sort out the truth from the tales of such early times.

Section 1 *Origins*

The Celts

The Celts were a very widespread people who lived (at one time or another) in a broad band right across Europe from Ireland in the west, through France, Belgium, Germany, Switzerland, Austria and as far as Turkey in the east. They were the tribes who opposed the landings of the Roman legions in south east England when the emperor Claudius decided to conquer Britain in 43 A.D.

This was not, however, the first meeting between Romans and Celts (or 'Gauls' as they appear in the Latin language). Some five centuries or so before the birth of Christ, tribes of warlike Celts in central Europe had been attracted across the Alps into northern Italy. The rich farmlands along the valley of the river Po were what drew them southward.

The Celts settled in northern Italy

A Celt touches the senator's beard

Eventually they came into contact with the Etruscans and drove them out. After a time the various tribes settled down on the plains of northern Italy. The Insubres had their capital at Milan, the Boii lived around Bologna, the Cenomani had tribal centres at Brescia and Verona, whilst the Lingones and Senones spread out along the Adriatic coast.

The latter group decided in the year 390 B.C. to raid even further south. A horde of them swept into Etruria and made their way down the Tiber valley.

Alarmed, the Romans sent an army to halt them. Unfortunately for Rome, her soldiers were badly beaten on the banks of the Allia river and the victorious Celts pushed on, unopposed. Three days after the battle, their vanguard appeared before the walls of Rome.

Many of the citizens had fled to other towns and cities – particularly to nearby Caere, leaving a handful of determined young men to garrison the Capitoline Hill. Brennus, the chief of the Celtic Senones, led his men almost unopposed into the rest of the deserted city.

In the Forum, they came upon a line of chairs made of ivory, upon which sat some of Rome's elderly senators. For a while the two groups stared at each other. Then a Celt put out a curious hand and touched a senator's white beard. The old man reacted angrily, so the Celt raised his iron sword and slew him. Soon all the senators were dead.

The Celts turned their attention to the Capitol which was still defended. Although the Celts were bold and daring, they were not keen on long, drawn-out sieges, but this is just what they were faced with. Their attempt to take the Capitol Hill went on for no less than seven months.

At one stage, the invaders tried to scale the cliffs at dead of night and the Romans were only saved by the sacred geese of Juno's temple. The birds cackled so loudly that the defenders were warned and sprang to their positions.

Eventually, the Celts were only persuaded to leave by being bribed with a large sum of gold. Incidentally, there is a legend which relates how a Roman army from one of the nearby cities turned up and drove off the Celts. This is almost certainly an invention, made up to wipe out the shame of the Celtic victory. The story goes on to talk about Rome being left 'a heap of smouldering ruins'.

The Celts assuredly did their share of destruction but it seems unlikely that the fires would still have been burning seven months after the arrival of the enemy. However, practically all the written records, including all mention of the first Roman laws, disappeared with the Celtic raids.

After the Celts had left, there were some Romans who wanted to abandon the city and move to the more easily defended city of Veii, which lay a few miles to the north west of Rome. They were overruled, although there were renewed attacks from old enemies such as the Volsci, the Aequi and the Etruscans.

The Romans rebuilt their city, reorganised their army and beat off their foes. The Celts came back again – several times, in fact, often allying themselves with Rome's ancient adversaries. For all practical purposes, though, the Romans had the last laugh, finally destroying their tormentors in Italy almost exactly a century after the first Celtic attacks.

Celts in the Alps in the 3rd century B.C.

Section 2 *Rome and her neighbours*

The Greeks

Seen through Greek eyes, early Rome was just another shepherds' village. Greeks had founded colonies all over the Mediterranean world, including southern Italy. Some of these settlements had themselves started colonies. It wasn't surprising therefore that Greek ideas should have been known to the pioneer inhabitants of Rome.

It was a two-way process; Greeks knew all about Rome by the fourth century B.C. and it was they who originated some of the legends about the birth of the city. They invented a man named 'Rhomus' after whom they said the city was called, and it was they who said that Romans were descended from Aeneas, the best known refugee from the Trojan war. It wasn't until the second century before Christ that Roman historians began to write down their city's story. When they did, they took the Greek legends and retold them in Latin.

Greek settlers in Italy were not ruled in any way by the cities in Greece from which they had come, but they were still Greeks and in their new homes they began to make a new life very similar to that which they had left behind.

As Rome expanded, it came more and more into contact with Greek ideas and thought. One practical result was the introduction by Greeks of grape and olive growing from their homeland to Italy.

Romans became aware of the fine temples these foreigners had put up to house their gods and in later years tried to imitate them. They also copied the architecture of the Greek dwelling house – at least, the better off Romans did. There were only one-roomed huts in the early days of Rome, but extra chambers were added to the 'atrium', as it was called, in imitation of the best Greek models. Rich Romans went further and included a central garden with fountains and statues and surrounded by even more rooms.

Other borrowings include the alphabet. Romans used it to write in Latin and it came not just from the Greeks but the Etruscans too. In fact, the alphabet was not a Greek invention but had been adapted from letters originally used in the Middle East. The very letter names 'alpha' and 'beta' which we think of as Greek are really near Eastern words for 'ox' and 'house'.

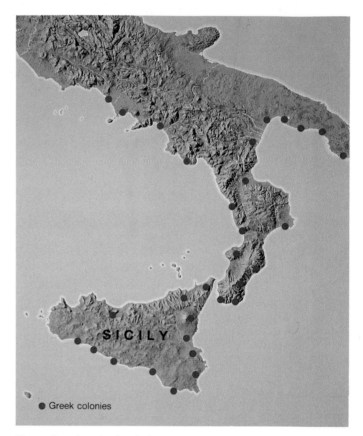

Magna Graecia – the Greeks in southern Italy

Romans cheerfully adopted Greek philosophy and tried to imitate the Greeks' skill as sculptors. Later Roman generals raided the homelands of Athens, Sparta and other city states in order to bear off to Rome countless thousands of original statues.

Latin laws often show a strong likeness to those of Athens. It was also to Athens that Romans owed their ideas of the use of coins to ease trade and commerce.

Greek colonies in southern Italy were known to the Romans as part of 'Magna Graecia' (Greater Greece), probably because in total area they were several times as large as Greece itself. The men of Rome had overrun the Greek Italian settlements by about 273 B.C. Later, in the next century, the legions were to conquer the Greek homelands themselves and thus to include Greece in the new and growing Roman Empire.

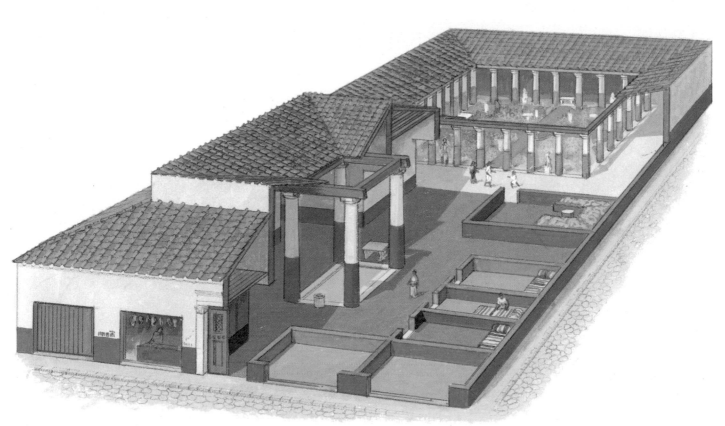

A Roman house built in the Greek style

Greek temple at Paestum, south of modern Naples

Carthage

Carthage was only a neighbour in the sense that the city lay just across the Mediterranean from Rome. She was not a neighbour in any friendly sense, being almost from the first a rival and merciless enemy. The people were known as 'Poeni' (the Latin for 'Phoenician') and anything to do with them as 'Punic'. We could ask a Roman historian to tell us about them and where they came from.

'They were a Middle Eastern people who settled on the coasts of what you call Lebanon about a thousand years or more before Rome itself was founded. Don't you call them Phoenicians?

'Just before Romulus started our own city, colonists from one of their seaports (a place named Tyre) began to build Carthage, or Qarthadasht as they called it in their own tongue. It means "the new capital". You may have heard of Dido and Aeneas, perhaps?'

We nod and he goes on, 'Dido was the sister of Pygmalion, King of Tyre. Her husband, the high priest Acerbas, quarrelled with the king who had him murdered. Dido was told in a dream to escape from

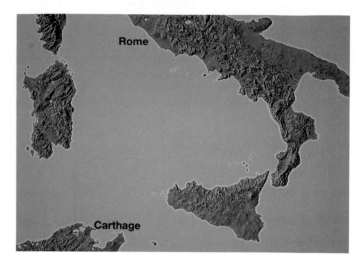

The location of Carthage in relation to Rome

Pygmalion. So with a handful of faithful friends and followers, she sailed away from Tyre and reached Africa where they began to build their new city.

'Now here comes a difficulty. Our poet, Virgil, says Dido fell in love with Aeneas, the ancestor of Romulus

Reconstruction of Carthage from the air

and Remus. This is a little confusing since the supposed "lovers" lived about three centuries apart. Still, it's only a story.

'Virgil goes on to say that Aeneas did not return her love and set sail for Italy. Dido, in despair, stabbed herself to death and her body was burned on a funeral pyre. Virgil may have a grain of truth here: there is a rumour that the Phoenicians sacrificed babies to their gods by burning them alive.'

'It's more than a rumour,' we say, 'for the burnt remains of hundreds of infants were dug up at the end of the last century. By the way, who were their gods?'

'Baal and Astarte ruled the Phoenician homelands but in Africa Baal became Baal Amon, or Moloch, and Tanith was the new name for Astarte.

'It was really a pity about Carthage. It had to be destroyed after it lost a series of wars against us but it must have been a fine city in its day.

'With nearly a million people, it was easily the largest of the many Phoenician settlements along the North African coast. It was founded by a nation of craftsmen, sailors and traders. It had a nearly ideal situation, being set on a rocky headland, protected both by large lakes on each side and triple city walls.

'These walls were a marvel of engineering. The outside one of the three walls was as thick as a tall man lying down and as high as seven tall men standing one on the other. The three walls ran across the headland for more than three miles. There was a defensive tower every seventy yards or so.

'The really clever part of the defences was the circular harbour, which was big enough to take two hundred warships at once. It was surrounded by another high wall so that an enemy could not tell from the sea whether there were ships there or not. There was a command post for an admiral on a little island in the middle of this lagoon.

'The place where the defenders could make a last stand was the citadel, a fortified hill more or less at the centre of the city. Houses were tall and built alongside narrow streets. It seemed a very difficult town to capture but we Romans not only took Carthage, we destroyed it.'

How they did it you can see in the next few pages.

A ship shed where ships were constructed and repaired

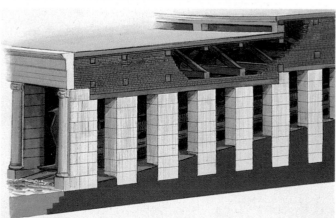

A sea fight

As Rome's soldiers conquered the nearby tribes and gradually extended her control over most of Italy, they came into contact with peoples whose homelands were not in Italy at all. Among these were the Greeks and the 'Poeni' from Carthage in North Africa.

The Carthaginians claimed the western half of the Mediterranean as theirs alone to trade in and resented Rome's interference in their affairs. The trouble, as far as Romans could see, was that the men of Carthage were experienced sailors, whilst the men of Rome were landsmen who knew nothing of the sea and its ways.

The two sides clashed over which one was to rule the islands around Italy, particularly Sicily. War broke out in 264 B.C. The men of Carthage must have thought that it was going to be easy to beat the Roman landlubbers. After all, the Romans not only had no warships, they were not considered skilled enough to make any. Then fate took a hand and a Carthaginian ship was driven ashore in Italy. Roman carpenters were sent to study it.

Within a few short weeks they managed to build and launch over a hundred *Roman* warships. While the actual hulls were being constructed, some benches and oars were made so that the men who were to row the vessels could get some kind of practice on dry land.

The idea was to use oarsmen to move the new ships as it was thought too difficult in the time to learn much about the management of sails. A favourite tactic used by the Phoenicians and other ship-owning nations was to try and ram the enemy vessel with the large armoured beak at the prow.

Because the Romans were good land fighters but knew nothing of warfare at sea, they thought up a new way of fighting a marine battle. As you can see from the picture, each Roman warship had a kind of drawbridge on its deck.

When an enemy vessel was within range, javelins were hurled at its crew and the two vessels were drawn together with grappling irons on the ends of lengths of rope. As soon as the two warships touched, the draw-bridge was allowed to fall. The far end dropped onto the enemy deck and a spike beneath it smashed through the planks, thus fastening both ships together.

Roman legionaries then ran across to the opposite deck and proceeded to fight just as though they were engaged in a land battle. Although this method worked well the huge boarding plank made the ship unstable and the technique was quietly dropped when Romans became more experienced at fighting battles at sea.

Hannibal and the Punic Wars

A young boy named Hannibal was taken to a temple by his father and made to swear on the altar of Moloch that he would be Rome's enemy until he died. When he grew up he became the general of the Punic army.

It was reported to him that a city in Spain which he had thought loyal to Carthage had since allied itself to Rome. He attacked the Spanish city and took it after an eight months' siege. This was more than long enough for the news to reach Rome. The government agreed that Hannibal had broken the peace treaty which Rome and Carthage had signed at the end of the first Punic War. Rome sent messages to Carthage demanding that Hannibal be handed over.

A Roman army was sent to Spain to take Hannibal but they arrived at his camp a few days after he and his men had left, heading eastward, no one knew whither. Guesses as to the size of his army varied from twenty to a hundred thousand but everyone agreed that he had with him at least thirty war elephants.

The Romans went back to Italy in ignorance of Hannibal's plans. The Carthaginian general had a very ambitious scheme in mind. He intended to attack Italy from its landward side, in spite of the Alpine ring of mountains which protects its northern frontier.

He led his army up the eastern coast of Spain, crossing the large rivers on rafts big enough to carry his elephants. He skirted the Pyrenees mountains, moved along the coast of France and over the river Rhône, making for the foothills of the Alps.

It was a tremendous achievement to get all his men, not to mention his elephants, up one side of the mountains and down the other   although many of the great beasts died in the attempt. Nowadays we can fly over the Alps, go through them by train in a tunnel, or drive over the top on well-made roads. In those days, the only routes were by way of goat tracks. They also had to contend with sharp rocks, steep slopes, narrow ledges, snowdrifts and ice even in midsummer and countless roaring mountain torrents.

Eventually, they arrived on the fertile plains of northern Italy, hoping that local tribesmen would come and join them. Few did, but the men of Carthage beat off the troops sent to stop them and began raiding towns and villages in the neighbourhood. There was panic in Rome and it seemed that no Roman armies could save the city from defeat. A supreme army commander

Rome and Carthage at the time of Hannibal

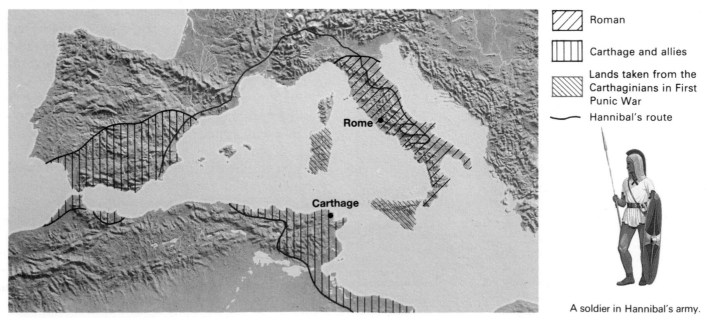

Roman

Carthage and allies

Lands taken from the Carthaginians in First Punic War

Hannibal's route

A soldier in Hannibal's army.

named Quintus Fabius Maximus was appointed.

Pitched battles were out of the question from then onward: the Romans were no match for the powerful African army. Fabius had learned from the first engagements that the legions had fought against Hannibal's men what would happen if he were to risk meeting the enemy face to face. If he lost, all was lost, and nothing could save Rome, so he did what he could to annoy and irritate the invader. Raids and ambushes were made on Carthaginian stragglers, retreats ordered when a battle looked likely, food stores and bridges destroyed.

This kind of thing went on for years until the Roman senate got fed up with the tactics of the 'Delayer', as they called him, and appointed another general to command. However, Fabius was proved right when the new man made up his mind to face Hannibal in a proper set battle. His legions were severely mauled.

Eventually the senate sent an army to Spain, largely ruled by Carthage at that time. The commander's name was Scipio. He was so successful he persuaded the senators to give him an army he could take to Carthage itself.

This new army beat a Carthaginian formation in Africa but their main aim was to draw Hannibal away from Italy. The trick worked and Hannibal faced Scipio at Zama near Carthage.

The Carthaginians were trounced and were forced not only to pay war damages but also to cut down the size of their fleet. Rome demanded that Hannibal be handed over to them. He escaped, but after years on the run from fear of Roman vengeance, he committed suicide.

Over fifty years later, the wars broke out again. Carthage lost again and this time refused the terms Rome offered, considering them much too harsh. It was all the excuse Rome needed. The city of Carthage was captured and burned to the ground. The few people left were sold into slavery. The charred ruins were smashed to pieces and the whole area ploughed up.

As a last gesture, orders came from Rome to sow salt on all the farms, so that nothing should ever grow there again.

Thus Rome gained control of the western Mediterranean and began to build an empire in earnest. The fruits of the Punic Wars were North Africa, Spain and the islands of Corsica, Sardinia and Sicily.

What Rome may have looked like

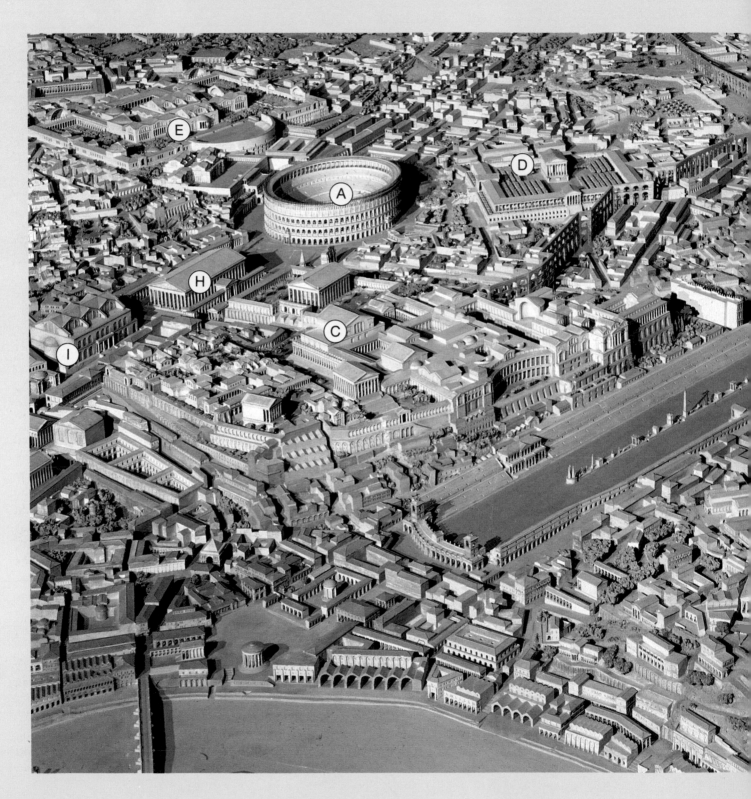

Photograph of a model of Rome

A   Colosseum
B   Circus Maximus
C   The Flavian Palace
D   The Temple of Claudius
E   Baths of Trajan
F   Aqueduct of Claudius
G   Baths of Caracalla
H   Temple of Venus and Roma
I   Basilica of Maxentius or Constantine   The Forum and Capitol are just off the picture to the left of the basilica of Maxentius
J   The walls of Aurelian

31

Aqueducts

When a village or town is fairly small, it can be supplied with food, fuel and water from its immediate neighbourhood. Things get difficult when the population increases sharply. No longer will a local stream or well yield all the water required. Not only did the number of Roman citizens increase rapidly, the custom of bathing every day also grew in popularity.

The city authorities decided to bring in water from the nearby hills. They did this by constructing aqueducts. Their engineers would find a suitable stream and make a shallow stone channel to divert its flow. To make sure the water got to the town, the aqueduct had to descend gradually over its entire length. This often meant that a roundabout route had to be chosen.

For instance, the very first aqueduct in Rome was built to the orders of Appius Claudius, in 312 B.C. The water source was only just over $6\frac{3}{4}$ miles away but the final length of the channel was more than ten miles!

Occasionally, the channel had to cross a ravine or a river valley. Then the water had to be taken across on a bridge. Some examples of this still survive. The best known is the Pont du Gard near Nîmes in France.

This was originally part of a twenty-five mile long aqueduct. There are some three hundred yards of it still to be seen. The method of construction was to make a row of huge stone arches, connect them with a flat stone bed, put a second and smaller line of arches on this bed and a third and final row on top of that. The water channel was at the very top. The arches were all solidly built: the ones at the bottom standing on foundations nearly thirty feet across. The middle arches are fifteen feet thick and the top ones ten feet wide. The overall height is about a hundred and sixty feet. A

Pont du Gard Roman aqueduct and bridge

An aqueduct outside Rome

Section of an aqueduct

similar one made of stone blocks but without mortar is still delivering water near Segovia in Spain.

In ancient times, there were eleven aqueducts bringing water to Rome. They had a total length of about three hundred miles and all but fifty of these were either underground or in covered ground-level channels. The stone coverings were to protect the water from insects, birds, small animals and evaporation from the hot sun.

The aqueducts usually ended in reservoirs in the city. Very few private houses had water laid on, although some illegal tapping took place. A civil servant sent by the emperor to investigate found that the five hundred or so slaves who looked after the system had started a business selling the unlawful right to be connected to any who could afford the bribes.

Water supplied public baths, lavatories and fountains, the latter being the source of drinking water for the poor who would collect what they needed in clay pots or goatskins and lug it up to their apartments, perhaps several floors above ground level.

The pressure was never great enough to supply any floor higher than ground level, even for those legally entitled to it. In fact, some parts of the city had to wait decades before they had a water supply at all. The usual reason for this was that some areas were too high: gravity-fed fountains would hardly work on hilltops.

Although the aqueducts delivered plenty of water, the lack of pressure meant that fires could only be fought with buckets of water – never a jet from a hose.

One curious aqueduct called the Alsietina didn't provide drinking water. All its liquid went to a special amphitheatre which could be flooded in order to stage water shows. These gradually came to a stop after Nero's time and the aqueduct was used to water gardens on the slopes of the Janiculum Hill.

Another specialised supply line brought both fresh and salt water from Ostia at the mouth of the Tiber to fill the fish ponds of the city markets. In this way, the Roman shopper could buy his sea food extremely fresh.

33

The baths

Romans were probably the cleanest of the early civilised peoples. We've seen how aqueducts brought water for fountains and baths – probably somewhere between fifty and two hundred gallons every day for each one of Rome's one million inhabitants.

Of course, the people didn't drink that much water: most of it went to supply the public baths. Many wealthy private houses in the country had their own suites of bathrooms – cleanliness was (from at least as early as the second century B.C.) fast becoming a passion with Romans.

By the time 'B.C.' had given way to 'A.D.', public baths were extremely popular. This was because most poor people lacked a bathroom. Even if they had been provided it would have meant a lot of trips up the stairs with jars of water – and where would the water have been heated?

It wasn't a great hardship to use the public establishments – there was nearly always one close by. In 33 B.C., Rome had about a hundred and seventy of them, but a couple of centuries later the number had grown to over a thousand. In any case, the use of the building and its hot water was either very cheap (less than $\frac{1}{2}$p!) or completely free. Upper-class Romans, seeking election to various posts, often paid the expenses of everyone for an entire year, hoping, no doubt, that such generosity would be rewarded with votes!

Most Roman towns, even small ones, had public baths. Volubilis in Morocco had two grand public baths as well as the private ones in the houses of those Romans who could afford them.

The interior decoration was often awe-inspiring – particularly to someone who had to live in a small flat with very little headroom. The baths provided by the

The baths of Caracalla. The domed building is the hot bath (caldarium). The cold room (frigidarium) is to the left. In between is the warm room (tepidarium)

The women's hot bath at Pompeii

emperors were huge. Those of Caracalla measured 750 feet by 380 feet – and that was only the main block. If you included all the walks, gardens and outbuildings, the whole complex covered about twenty-seven acres (equal to about fifteen full-sized hockey or football pitches) and these were not the largest.

The visitor would notice the ceilings a hundred feet above his head and the richness of the decoration. Walls were set with rare stones, corridors lined with marble pillars, whilst floors were often patterned with mosaic designs in pieces of coloured tile called tesserae.

Perhaps we could ask one of the young bathers to tell us what a Roman bath is like.

'My name is Junius. I'm twelve years old and I'm going to bathe on my way home from morning school.'

'Are those your towel and bathing trunks you're carrying?'

'My towel, yes – although you can hire one if you want. But we don't wear bathing costumes. You don't when you take a bath, do you?

'We wait for the opening bell to sound: it signals the end of the women's and girls' time in the baths. Then we go in through one of the four main entrances to the undressing room. It's a good idea to have someone guard your clothes – you might get them stolen otherwise. After that, you can do all kinds of things: you

might go to the hot, dry rooms (we call them "Spartan" baths) –'

'Saunas,' we murmur.

'You sweat a lot and then dive in a cold pool. Or you go to the open air spaces where you can wrestle or play ball games with your friends. Otherwise you work up a sweat by going through the frigid and tepid rooms to the hot steam room –'

'Turkish baths,' we mutter.

'–where you treat your skin to an oiling followed by a scrape down with a strigil. You can hire a slave to do it for you but that's expensive. So is being massaged and perfumed.

'If you're rich, you can afford all these treatments but if you're like me, you splash yourself with hot water from the basin and scrape yourself down before diving into the cold bath or going back through the cooler rooms again.

'You could, if you like, use some of the other attractions. There's a library, laundry, reading-rooms, gardens, tailor's shop, barber's, manicurist, rest-rooms, wine shop, restaurant and gymnasia.

'Most Romans bathe every day – some more than once. One of our emperors, Commodus, is supposed to have bathed no fewer than eight times every day!'

Temples

'I understand you want to look round some of our temples?' We murmur our agreement and the speaker says, 'I'm Lucius Brennius and I'd be happy to show you round. This is our local temple, of course. It's quite old – two or three hundred years, I think.'

'It's very like a Greek one.'

'Yes, but there are differences. To start with, the stone platform on which it stands is much bigger. Greek temples are two or three feet above ground level and you can get on to the platform from anywhere around it. Now, a Roman temple base is often nine or ten feet high and you have to use the staircase at the front to get up.'

'That's right – and your local temple doesn't have columns all round it – only a couple of rows at the front and just a line of half columns down the sides. Are there any at the back?'

'No, they aren't needed. Let's go up the stairs. This is the open part of the temple. At the back is the interior room, or cella. We can't go in, as it's locked. It's where we keep the statue of the god to whom the temple is dedicated.'

'And this one is – ?'

'It's the house of Mars.'

'What's the building made of?'

'Concrete, stone – things like that.'

'How do Romans make concrete?'

'You take a lot of broken bits of brick or stone, mix them with sand or gravel, adding cement and water. You stir them all together, spread the mixture out wherever you want it and wait for it to set like a rock. With wooden shuttering you can make an arch out of concrete. A long arch gives you a vault. Turn the arch round on its centre spot and you've got a dome. There's a temple in Rome called the Pantheon which has an ordinary rectangular entrance but a round building behind it, surmounted by a huge dome. Hadrian rebuilt it from an earlier model.

'The dome is almost a hundred and fifty feet in diameter and it sits on walls twenty feet thick. There's a circular hole twenty-seven feet across at the top to let the light in. Of course, that's not the only round temple, although they aren't as common as the oblong ones.'

A typical Roman temple

'What does "Pantheon" mean?'

' "All the gods." Usually a temple is the house of just one god but even apart from the Pantheon there are a few places dedicated to two or three different heavenly beings.'

'You didn't finish telling us about what the buildings are made of.'

'Well, the early ones were carpentered from tree trunks but we usually use mined materials now – you know, whatever's available on the spot, perhaps granite, limestone or volcanic tufa. That's a Roman speciality. It's a dull stone but it can be hidden under a coat of plaster or better still enveloped in white marble. A lot of temples are improved with a nice light-coloured paint or decorated with rare stones.

'Some have strong rooms as well. The temple of Saturn in Rome is where the state treasury is situated.'

'That's surprising.'

'Not really. A few temples have large amounts of gold entrusted to them: they even lend money at interest or change cash for foreigners – banks, I suppose you'd call them. In Pompeii and some other towns, the temple is where you find the weights and measures office. There's a slab of stone with scooped out hollows for measuring corn, oil and wine.

'But the chief purpose of the temple is religion. It's the house of the god. I can tell you more about the actual gods, if you like.'

The Pantheon

'We haven't time now,' we tell Lucius, 'but thank you all the same.' However, if you want to find out about who the Romans worshipped, turn to p.72.

Private houses and flats

As in the matter of temples, the better kind of Roman house was copied from a Greek original. But in the beginning, things were very different. The dwellings of the pioneer shepherds on the seven hills were little more than one-roomed huts with a hole in the roof to let the smoke out. As time went on and people became somewhat better off, they added a room or two here and there, around the original hut.

The newer houses followed the same pattern and the 'hole in the roof' room changed into a sort of entrance hall, or 'atrium', as they called it. Below the opening was a shallow trough to collect rainwater. This arrangement continued even into the wealthy times. A rich Roman's house still had its atrium and rainwater pool. Several rooms opened off the atrium – mostly bedrooms.

At the far end were reception rooms and beyond them, a small garden perhaps with statues and a fountain. There was a roofed colonnade running round the garden. Nearby was the kitchen and dining-room. If possible, a bath suite was included.

Walls were plastered and painted with scenes of the countryside, including birds and flowers. The bright colours of the paints were echoed by the different hues of the mosaic floors. Mosaics were made from thousands of pieces of coloured stones or tiles set out in pictures or geometric patterns.

A very large house in Rome might occupy all the space bounded by four streets, thus forming a block, or 'insula' (island), as they called it. Partly to bring in more money and partly to insulate the family from the noise and bustle of the streets, several street-side areas were let off to shopkeepers.

A slave might be stationed at the front door to keep out unwanted callers. A house pet could be chained up nearby to frighten off burglars and similar lawless ones. A mosaic by a street door in Pompeii has a picture of a dog, with the words 'Cave Canem' ('Beware of the Dog') beneath it.

Wealthy houses like these were pretty rare even in Rome. Most people were poor and were forced to live in tenements if they wanted to remain in the capital city. Land was scarce, so landlords had nowhere to expand

Block of flats (insula) at Ostia

but upwards. Because greedy owners often used shoddy materials for building, the danger of collapse increased with the height of the building. This caused various Roman emperors to make laws forbidding blocks of flats over a certain height – for example, sixty feet. Even at that, the builders could still get in as many as eight or nine floors. The greater the number of floors, the more extra tenants could be squeezed in and the higher the profits of the block owner.

There were no lifts, of course – nor could poor people afford expensive glazed windows. You didn't need them

Wall paintings from the house of the Vetti brothers at Pompeii

Insula at Pompeii

in a sticky Roman summer but when the weather got worse, the only way to keep the flat's temperature up and the rain out was to close the wooden shutters.

As this plunged the room into darkness, the flat dwellers had to make do with smelly olive-oil lamps. Heating was done by means of a portable metal brazier which burnt charcoal. Rich people had a safer and more efficient system. Main rooms were built over a hypocaust which allowed hot air from an outside furnace to circulate under the floor.

In the tenements, accidents with heaters and lamps were common and – as the apartments were largely made of timber – extremely dangerous. Not only the flat where the brazier accident had occurred was at risk: the whole block could go up in flames. If this happened there wasn't much chance of saving it. There was a sort of fire brigade system under the emperors but there were no mechanical pumps or hoses. All the firemen could do was to form a human chain, passing leather buckets from hand to hand and hoping the water delivered to the block was enough to quench the flames. If this didn't work, they had to try to pull down the building with hooks on long poles, to stop the fire spreading.

It has already been mentioned that there was no water laid on in the blocks. Not only did the flat dwellers have to carry up every drop of water they wanted, but there were no internal lavatories. The tenants were lucky if there was a public lavatory fairly near!

Shops

In the first few years after the founding of Rome, the shepherds and farmers were self-sufficient peasants. If you don't want much beyond the roughest food, shelter and clothing with the crudest of pottery and carpentry, you can do it all for yourself. As soon as the country-men and their families began to hanker after something better, their standard of living could only rise if there were craftsmen spending most of their time on the special thing they were good at.

A man might make better baskets, spades or shoes than anyone else in his neighbourhood. The neighbours would prefer his products to those they could make themselves and they'd pay him – at first with farm produce and later with coins.

It wasn't long before open areas in the valleys between the seven hills of Rome were used by those with something to sell – either what they had grown or what they had made. Dozens of small stalls were set up at which people could buy whatever took their fancy.

After many years, specially built shops were provided. These might be in rows along a street or the side of a square. Other locations were in odd corners of a rich man's house or block of flats. Glass was known in Roman times but it was only used for fancy containers, drinking vessels or the rare windows in well-to-do houses. It couldn't be made into very large sheets and would have been much too expensive for shop fronts, say, in any case.

For this reason shops were closed at night with wooden shutters and totally open during business hours. In many cases, the man of the family made the trade goods in a workroom behind the shop. His wife and grown-up children sold the goods to the public and the whole family lived in an upper room or an apartment behind the workroom.

Because transport costs were so heavy, there was never any chance of nationwide firms of chain stores with outlets in every large town. So local craftsmen made their wares of local materials to local designs. You wouldn't expect to match the pattern on a favourite piece of cloth in any town but your own. The only goods that seem to have been made and exported on a large scale were some items of pottery.

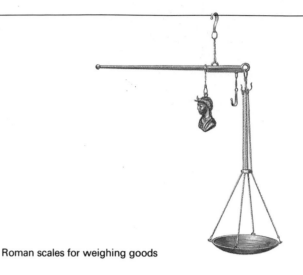

Roman scales for weighing goods

The stalls in the forum, or open square, were where the male slaves did the shopping. Women, particularly rich ones, rarely shopped except for cosmetics, clothes or jewellery.

Market stalls provided meat, fish, vegetables and fruit. The fish was probably dried – if you wanted fresh stuff you went to a proper shop where you could choose your fish live from a tank of water. Other, and more permanent, shops sold shoes, knives, ironware, rope, leather goods, poultry, wine, bread and many more items. Rome, like any large modern city, was a place

Plaster cast of shuttering and door of shop

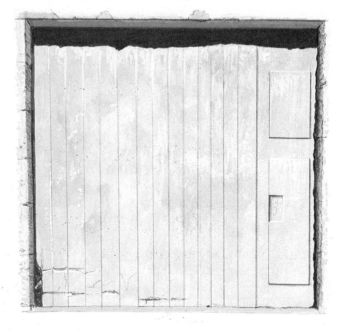

Street scene

where you could buy almost anything.

One service provided, which we don't have anymore, was the public oven. Because of those who lived in flats with nowhere to cook, the baker hired out his ovens. Poor people brought him their dinners and he cooked them for a small fee.

Permanent shops commonly had a name painted outside. It wasn't the name of the owner but that of the product he sold, or the service he provided, e.g. 'oil', 'books', 'shoe repairs' or 'barbers'. If there weren't any words, the shop owner probably had a sign – for instance, tavern keepers draped their doorways with green boughs.

Goods were sold in standard measures and the 'steelyard' type of scale weighed what customers had bought. In earlier days, when things were simpler, a Roman got what he or she wanted by swapping things. It must have made trade a lot easier when (just after 300 B.C.) lumps of copper and bronze came into use to pay for purchases.

Some time before 200 B.C., round metal coins made their appearance – mostly gold and silver for trading abroad and small change of copper, bronze and brass for local deals.

Even the smallest of Roman towns had its share of shops. When St Albans in Hertfordshire was being excavated a few years ago, the diggers found the basements of a row of shops near the open air theatre. It isn't too hard to imagine slaves buying olive-oil, wine or ready-cooked food from the shop assistants or apprentices and paying with bronze coins here in England, at the furthest boundary of the empire.

Nero's Golden House

Such was the method of government in ancient Rome that power was eventually entrusted to one man alone. The emperor, as he was called, had virtually the power of life and death over all of his subjects and the authority to have his every whim carried out exactly.

From the time of Augustus, some of Rome's early emperors lived in houses little different from those of their well-to-do subjects, but as time went on their palaces got larger and more elaborate. Many of these early royal residences were on the Palatine Hill – hence the word 'palace'.

Things were not so bad just as long as the emperor was a fairly sane and responsible person. Unfortunately, not all Roman emperors were reasonable: some were a little odd and others not far from insanity.

Nero came somewhere near the insane end of this scale. He didn't behave in a civilised way and many murders and executions were his responsibility. The same charges could have been levelled at other rulers, but Nero's murders included those of his wife and his own mother.

However, the thing which caused his standing to decline with his more sober and dignified subjects was taking part in public song and poetry recitals. He entered competitions and the organisers dared not let anyone else win.

It was known that Nero wanted a large area of land on which to build a palace, so when a considerable part of Rome caught fire and cleared the region in question, people began to say that perhaps he must have been responsible. Although Nero was not in the city at the time, there was a certain amount of evidence that some of his servants had been seen at the place where the fire had started with flaming torches in their hands. To divert the people's suspicion he accused the Christians and had them tortured and executed by the thousand.

He then took over the burnt-out parts of the city and began to build himself a vast palace. It was called 'the House of Passage' as it led from one hill to another, so extensive was its lay-out. Unluckily for the emperor, it too burned down almost as soon as it was finished. The people groaned, for this meant more taxes to be paid in order that the great work should be redone. They were

right – it was rebuilt and retitled the 'Domus Aurea' ('Golden House').

The Golden House was not one building but several and the whole estate occupied an area of almost a square mile in the very middle of one of the most crowded cities of all time.

Nero had a lake dug out where the Colosseum now stands and surrounded it with parkland, farms, forests and vineyards. Dotted about were rich and elaborate pavilions, some connected by colonnaded paths. Statues adorned the grounds – a few from local sculptors but most taken from Greek cities and temples. On a

single expedition, Nero's servants sent home almost two thousand statues from conquered Greece.

After many years, the people of Rome could stand the behaviour of their murderous, half-mad ruler no longer. The senate turned against him and to avoid punishment and shame, he committed suicide.

After his death, the parklands were broken up and built over. Not only houses and blocks of flats appeared but also many public buildings such as baths and temples. Among them was the Flavian amphitheatre, named from Titus Flavius Vespasian who had it built. It stands where Nero's ornamental lake once was and near where the late emperor's statue once stood. Because of its huge size, Nero's statue was known as the 'colossal' statue. From this word, later generations began to speak of the arena as the 'Colosseum'.

Curiously, the Golden House itself was burnt down some forty years after the emperor's death. Later rulers tried to blot out the memory of this near madman and erected their baths and temples over the remains of his 'folly'. In many cases, these same baths and temples have themselves collapsed into ruin, leaving the buried remains of the Domus Aurea to be found by future archaeologists.

The visitor to today's Rome may be able to see what's left of the splendour – some bits and pieces still remain – for example some sections of decorated wall. There is a high vaulted corridor with painted false windows, through which one can apparently see distant vistas of beautiful landscape.

Alas! there are no longer any traces of the pipes which sprinkled Nero's guests with perfume, nor the ivory ceiling through which flower petals sifted slowly down on the diners. And although there is an octagonal room rather like a small Pantheon, there is nothing left of the cooling system said to have been used in hot weather, i.e. streams of cold water running down the stairs and out through drains in the floor.

Nero's Golden House with the Temple of Claudius and Aqua Claudia (aqueduct) in the foreground

Why the kingdom ended

Tullia rides her chariot over her father, Servius Tullius

You could get more than one answer to the question, 'Why did the kingdom end?' The average Roman of about 35 B.C. would tend to believe the legends, whilst a modern historian would have another and more down-to-earth answer. Let's ask both of them what they think: we'll start with Gaius, a citizen of Rome.

'I *know* why we kicked out the kings,' says Gaius, 'I don't have to guess. I suppose you know what our last ruler was called?'

'Tarquinius Superbus,' the historian says, 'Tarquin the Proud.'

'That's right. I believe you've already heard that he behaved so badly as king that he was thrown out? Well, his method of gaining the throne should have warned my ancestors what he was like.

'He married Tullia, the daughter of the king, Servius Tullius, to give himself a claim to the throne when his father-in-law died. But Tarquinius couldn't wait for the king's natural end – he went down to the senate dressed in royal robes, threw the old man down the stairs and took his place on the throne. He sent men to finish him off but Tullia made sure of her father's death by running him over with her chariot.

'From then on, the new king had to have a bodyguard – there were still many Romans loyal to the old king's memory.'

'They hated him because he was an Etruscan,' says the modern historian.

'Perhaps, but he was usually successful in war and gained supporters because of that.'

'What about his sly behaviour with the Gabii?'

'Oh yes. He sent one of his sons, Sextus, to the town of Gabii, one of Rome's neighbours. He pretended he was frightened of his father and wormed his way into their confidence. They gave him command of a small group of soldiers and he led them against a Roman detachment. He waited years but as soon as he was appointed supreme general of all their forces, he rewarded them by surrendering them all to Rome. It was, of course, the plan from the start. None of that family had any second thoughts about breaking promises. No, they were all sly, that lot.

'Take what happened when the king sent two of his sons and their cousin to Delphi to ask the meaning of a vision he had had. The young men decided instead to ask the oracle which of them should reign when their father died. The oracle's answer was "Which ever of you kisses his mother first." Few people heard what the cousin said on their return to Rome. He pretended to stumble full length on the ground, muttering as his lips brushed the soil, "The earth is our mother!" The sly devil was trying to put himself forward as the next king.

Etruscan coin

What a tribe!

'Unfortunately for his chances, the wife of another cousin, a woman named Lucretia, complained that Sextus had, with the king's approval, forced his unwelcome attentions on her. She told all Rome what she had suffered and killed herself in public –'

'I know,' the historian interrupts, 'that what you've said is what is supposed to have made Roman citizens drive the Tarquin family out, but there is another reason.

'The last three kings had all been Etruscans – foreigners as far as Rome was concerned. Etruscans were Rome's biggest rivals in Italy and it seemed that they were about to dominate Rome and other Latin cities by force. They had already taken over much of the trade in northern and central Italy. In any case, I'm sure Romans resented Etruscan culture and success – even in spite of the fact that they had borrowed some of Etruria's best ideas – for instance writing and the use of coins.

'No, Romans were jealous of the Etruscans and their superior ways. *That's* why they were kicked out and why Romans vowed never to have another king. Why, they even passed a law saying that any man who as much as talked about having a king back should be executed! Centuries later, Julius Caesar was murdered – not because he wanted to be king but because he was *suspected* of wanting it!'

The republic and its end

To run the government, Romans elected two consuls who could rule for one year only. Both had to agree before anything at all could be done. It only needed one consul to say 'Veto' ('I forbid') and the matter was dropped.

Other officials were elected – almost entirely from the ranks of the richer citizens. These were the descendants of the first settlers who had had the time to acquire land and become wealthy.

One drawback of the Roman system was that poor people scarcely got a look in. The patricians, as the rich citizens were called, provided all of the three hundred or so senators who ran the affairs of the republic. They were also the group from which came all the priests, army officers and senior civil servants.

The plebeians, as the poor people were called, fought hard throughout the republican period to get on equal terms with the patricians.

They never quite managed it and their struggle was complicated by various military governors who, seeking to restore public order, thought that what Rome needed was to have a single person in charge – each one thinking of himself.

At the time when the Etruscans were expelled, Rome ruled about three hundred and fifty square miles of Italy: just over two and a half centuries later, the figure had risen to ten thousand square miles. Overseas and other territory was added as the result of wars against the Phoenicians of North Africa and Spain; against the Greeks and Macedonians of south east Europe; and against the Gauls, or Celts in northern Italy and southern France.

A number of military adventurers were responsible for extending Rome's frontiers but their ambition to be supreme ruler threatened the peace and order of the republic. Among these men were Sulla, Pompey and Julius Caesar.

Julius Caesar conquered Gaul and raided Britain but he showed no sign of giving up his military dictatorship. A group of conspirators, among them Brutus and Cassius, were afraid that he wanted to be king. Romans had had their fill of unpopular kings in the city's early days, so they murdered him.

Assassination of Julius Caesar beneath Pompey's statue

Outstanding men were often given a division of fighting legionaries to do some special job such as the defence of a frontier, the conquering of a new province or the clearing out of pirate ports. Trouble often followed when such men refused to give up command and used their position to try and take over the government of the empire, as it was rapidly becoming.

Rioting and civil war followed Julius Caesar's murder. On one side were the murderers and the armies they controlled and on the other were the armies of Julius Caesar's great-nephew, Octavian, together with Mark Antony and Lepidus. These latter three ruled Rome together just as soon as they had beaten the armies of the murderers.

They were known as the 'Triumvirate' (Latin for 'three man rule') but it didn't last very long. Lepidus rebelled against Octavian and spent the rest of his life in exile. Mark Antony fell in love with Cleopatra, the last pharaoh of Egypt, and tried to set up a rival empire. His army was badly beaten by Octavian's at the battle of Actium in 31 B.C. Both Cleopatra and Antony killed themselves, leaving Octavian to rule Rome alone.

Although he protested that he was merely restoring the republic, he was in fact the very first emperor.

The republic was at an end.

Pompey, who was defeated by his rival, Julius Caesar

Caesar's troops landing in Britain, 55 B.C.

The early emperors

A Roman triumph

Augustus Caesar. This is the name by which Octavian is known to history. 'Caesar' was his adopted family name and 'Augustus' ('the dignified one') was the title given to him by the senate. He was careful not to seem like an emperor, pretending he was merely 'the first among equals'. The elections of men to the senate and to other public offices were allowed to continue, thus persuading the people that they were still in control, but only those candidates approved by Augustus were permitted to stand.

Augustus himself refused the title 'rex' ('king'), preferring to be called 'commander' or 'chief'. Some people were not fooled but he did bring peace for many years and they were willing to take him as he was because he had ended the decades of unrest and civil war.

Augustus Caesar

He reigned from 31 B.C. and was nearly eighty when he died in 14 A.D.

Tiberius. He was the stepson of Augustus. At the start of his reign he continued the slow, steady progress begun by Augustus. His very slowness was taken by some to be stupidity and several plots against him were discovered. He felt he was being persecuted and retired to the Isle of Capri, leaving Sejanus, the captain of the guard, to run the country.

To justify himself, Sejanus found plots against his master where none existed and many innocent people were executed. He even schemed to get the throne for himself but his behaviour was so suspicious that he himself was arrested and put to death.

Tiberius was almost as old as Augustus had been when he died in 37 A.D. To succeed him, he had chosen a great-nephew named Gaius.

Caligula (37–41). This was Gaius's nickname: it meant 'little army boot', a title he was given as a child because he liked to dress up in a toy soldier's uniform.

The beginning of his reign was uneventful: he was welcomed as a pleasant change from the gloomy and suspicious Tiberius. Unfortunately he had an illness shortly after becoming emperor. It affected his mind and his behaviour became more and more outrageous. Anyone who showed the faintest disagreement with him might find himself thrown to the lions in the arena. He forced the senators to elect his horse consul!

No one knew who would be Caligula's next victim, so it was with relief that they heard he had been murdered. He was the first emperor *known* to have died by violence (although some historians think that both Augustus and Tiberius may have been murdered). However, Caligula was not the last emperor to be assassinated by a long, long way!

Claudius (41–54). The next emperor was the uncle

Caligula

Nero

of Caligula. He had only preserved his life during his nephew's reign by pretending to be a stuttering idiot and therefore harmless. It's possible that he began the successful invasion of Britain to prove he wasn't cowardly or stupid. He was in Britain for just a fortnight in 43 A.D. but Roman legions were here for another four centuries.

Claudius was unlucky enough to meet and marry Agrippina, who nursed a secret ambition to rule Rome herself. As wife of the emperor this wasn't really possible, so she persuaded him to adopt her son by a previous marriage as the next emperor and to cut his own son out of the succession. Agrippina then poisoned Claudius, reasoning that her son as emperor would do as his mother told him. The son's name was Nero.

Nero (54–68). He did as he was bid – at least, to start with – but then began to get impatient. When Agrippina refused to let him divorce his wife, he had his wife murdered. Then to make sure that his throne would not be taken from him, he had Claudius's disinherited son murdered too. Now nursing a bitter hatred of his mother, he made her travel on a special boat which was designed to collapse when it was at sea. Agrippina swam ashore. Later Nero sent his thugs to assassinate her.

We've seen how he was suspected of having set fire to Rome and how his sprawling palaces were resented by the ordinary Roman. His pretensions to be a poet and singer annoyed a lot more people – not least those who found they were competing against him and knowing there could only be a *royal* winner.

Eventually, a number of his military commanders revolted and one of them, named Galba, was made emperor in his place. When Nero heard that he had been sentenced to death by whipping, he took his own life.

The empire expands

At the time of Augustus the Roman empire was almost at its greatest extent. The final conquests took place during the reigns of the five emperors who followed Nero. Maybe this elderly gentleman can tell us something about these five emperors.

'I certainly can,' says the man with the white hair. 'My name is Decius Marullus. I'm eighty years old and I've lived through the reigns of more than five emperors. I was born in Nero's time and now we have Hadrian on the throne.'

'Who were these emperors?'

'After Nero there was a time of confusion when no less than four military commanders raced one another to Rome to take his place. Galba was proclaimed but there was fighting even in Rome itself between the rivals – Galba, Otho, Vitellius and Vespasian.'

'Wasn't Vespasian a legionary commander in the invasion of Britain?'

'That's the man. It was he who came out on top. It was strange really – Nero had favoured his army career and even sent him to Judaea to run the war against the Hebrews. He'd done this because Vespasian wasn't a patrician. Nero thought that no plebeian could ever get to be emperor, so there was no danger in promoting this common soldier.

'Vespasian tried to rule as Augustus had done and to behave as a fair and reasonable person. This was a pleasant change from the madman whom he'd replaced.

'He'd beaten the Hebrews with the help of his son, Titus, so Vespasian chose Titus to succeed him. The emperor gave orders that the conquest of Britain should continue.

Trajan's army crosses the River Danube into Dacia

50

A scene from Trajan's column – burning a Dacian town

'At home, Vespasian had to increase taxes, as Nero had almost emptied the official money chests.

'Rome was improved with the provision of many new buildings. There were temples and baths of course: there was also a new public square and the foundations of a huge amphitheatre that was to rise on what had once been the bed of Nero's ornamental lake.

'Titus "assumed the purple", as they say, after his father's ten-year reign but he ruled only for two years. During those two years, there was another disastrous fire in Rome and an eruption from the volcano called Vesuvius which buried several towns and villages, including Pompeii and Herculaneum.

'When Titus died, Domitian took his brother's place. During his reign, the conquest of England was completed and the legions moved into southern Scotland. The emperor himself led his troops against the Dacians in eastern Europe. On his return to Rome he rode in triumph through the city. Unkind critics said that he hadn't beaten the barbarians at all but paid them to go away. They accused him of dressing slaves in barbarian costumes and pretending they were Dacian captives.

'He had some success with the opening up of trade routes to India and China but was less lucky in dealing with a "wine lake". The enlargement of the empire had meant that wine from North Africa, France and Spain could now compete with the Italian vintages in Rome, and it flooded in. Laws were passed to limit production – some vines had to be dug up and not replanted – but the laws didn't really work.

'Domitian wanted to be worshipped as a god but many Jews and Christians refused, even at the risk of being thrown to the lions. The emperor was so cruel to anyone merely suspected of plotting against him that sooner or later someone was going to follow a well established pattern and murder him. Those who had planned his assassination then picked an old lawyer named Nerva to be the next ruler.

'He reigned only two years and is mostly remembered for a scheme to look after poor orphans. He had time to nominate Trajan as his successor before he died.

'Trajan was from Spain and reigned nineteen years. He was the first non-Italian to be emperor. He came from a military family so I suppose it's fitting that he should be the one to make the last additions to the empire.

'Unlike Domitian, he really did beat back the Dacians. He reorganised the provinces along the Rhine and added Armenia, Assyria, part of Arabia and Mesopotamia to the empire.

'He also ordered the building of squares, libraries, memorial arches, baths, statues, public halls and theatres, not just in the capital but all over the empire. Here in Rome he had the story of his Dacian campaign carved in a long spiral strip around a stone column!'

51

The decline of the glory

Our informant, Decius Marullus, who told us about the growing empire on the last page, was living in the reign of Hadrian. This was the emperor who thought it more important for Rome to guard and secure a somewhat smaller empire than that which Trajan had left him.

It wasn't that he was more timid than the last emperor. The fact was that Trajan's adventures had wasted more men and money than Rome could afford. Attacks by barbarians convinced Hadrian that what was needed was not more empire but better defences.

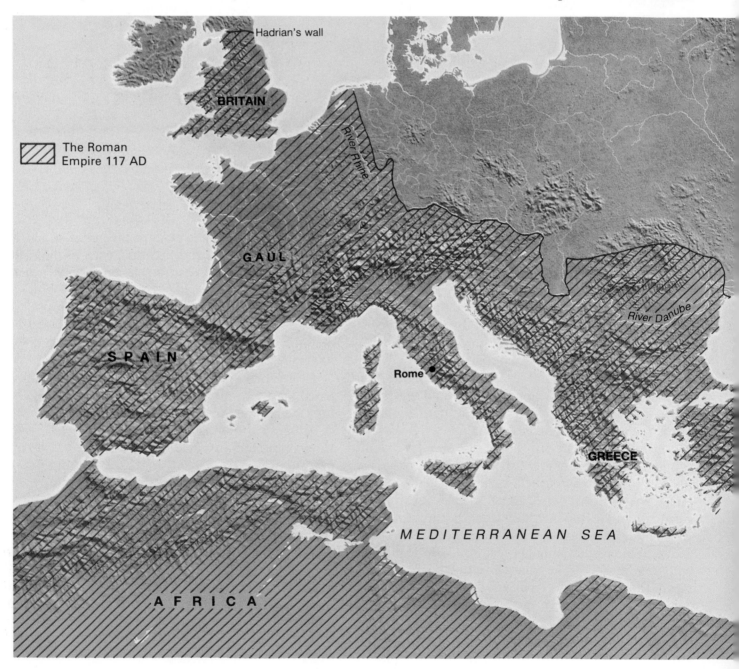

To that end, he abandoned Trajan's conquests in Asia Minor and supervised the setting up of lines of fortresses between the upper Rhine and Danube rivers.

Some five years after he had become emperor he arrived in Britain and ordered the building of a wall from near modern Newcastle right across the country to near modern Carlisle.

Trajan's empire was the largest it was ever to be: from the time of Hadrian it was a struggle to keep things as they were. Later emperors either fought to try and hold on to what they had or for various reasons had to watch while pieces of the empire crumbled away.

The Roman Empire at the beginning of Hadrian's reign, 117 A.D.

ASIA

Antoninus Pius, although a soldier, did very little with the army after he came to the throne, preferring to concern himself with domestic affairs. It's true that his twenty-three-year reign was a better time for the majority of Romans than most other periods of the city's history. All the same, the strain of running such a huge organisation had grown too much for one man to manage.

Antoninus adopted a young man named Marcus Aurelius and gave him the title 'Caesar', keeping the word 'Augustus' for himself. Together they ran the empire and when Antoninus died in 161, Marcus took his place.

Marcus Aurelius would have been far happier browsing in a well-stocked library. Instead of reading, thinking and writing, he was burdened by the cares of an empire he had never sought to rule. A good deal of his time was spent riding the frontiers, pushing back the waves of barbarian invaders, who were themselves being forced towards the empire's borders by wild tribes even further away.

On top of these troubles, there were several epidemics of plague which ran through the native populations like wildfire. Often, the diseases were being brought into the empire by legionaries returning from Asia Minor.

It seemed to Marcus that he was like a swimmer, who, far from moving forward, was energetically treading water to stop himself sinking and drowning. He was glad of the help of his only son, Commodus, who succeeded him as emperor when he finally died, absolutely worn out.

Commodus was as unlike his father as can be imagined. Marcus was thoughtful, kind, educated and well mannered: his son was a selfish, coarse brute who really enjoyed the beast and gladiator fights. He loved to see the blood flow and often boasted that he could have fought better himself.

Soon, idle boasts were translated into action as the emperor decided to take part in the fights himself. To make sure his sacred person was not injured and that the emperor could not lose, his opponents were given blunt weapons made of soft metals such as lead.

His intimate friends were gladiators, animal trainers, boxers and all kinds of people unsuited to court life. When plots against him were uncovered, he couldn't be bothered to hold a long enquiry, followed by proper trials, he merely ordered all the suspects to be executed.

Finally he was murdered by one of his own guards – a fate he shared with quite a few other Roman emperors.

Before we take the story too far, it would be as well to find out a little more of what life was like for ordinary people.

Section 5 *Daily life*

Pompeii

A great deal of our knowledge of daily life in Roman times comes from Pompeii and Herculaneum, two Roman towns near the modern city of Naples, on Italy's south-west coast.

You may remember that Vesuvius had erupted in 79 A.D. No one seems to have been prepared for the devastation, in spite of a previous outburst from the volcano in 63 A.D. Most people were apparently going about their business in August 79 when the first rumbles were heard.

Earthquakes were not uncommon in those parts but these seemed stronger than usual. On the 23rd of that month Vesuvius blew up. The top of the mountain developed a huge cloud shot through with flames. The

The destruction of Pompeii, August 79 A.D.

Plaster casts of a dog, a man, and a loaf of bread found at Pompeii

Wall painting of a religious ceremony from Pompeii

sea drew back from the land, leaving fish flapping on the wet sand. There was darkness during the day time and ash began to fall from the skies.

Some escaped, protecting themselves from the hot ashes by tying pillows over their heads and cutting out the sulphurous smell by winding a scarf over nose and mouth. Others were not so lucky and didn't manage to get away as more lethal gases spilled into the air. The few that were left of Pompeii's 40,000 inhabitants just dropped where they were and died.

The ashes continued drifting down and the buildings slowly disappeared under layers of the stuff, which gradually hardened until it was almost as hard as rock.

The ruins then lay buried and forgotten until the eighteenth century. Then a small group of people interested in history began to uncover the streets, fountains, public buildings and private houses of a Roman provincial town. People have been digging ever since and the town has still not been fully unearthed.

So much has come to light, though, that we can get a new look at the everyday life of ordinary Romans. What did the archaeologists discover?

The remains of countless wall fragments were painted with flowers, fruit, animals, patterns and above all, Pompeian citizens going about their affairs. Pompeii has provided the bulk of all Roman painting known to the experts.

From the foundations and lower wall portions, historians can work out the most popular arrangements for private houses. They can also see that there were understreet sewers, sometimes with inspection covers. The streets they drained were very narrow by modern standards – only about twelve to twenty feet wide, sometimes with wheel ruts in the stone and occasional side pavements, slightly raised and topped with asphalt.

There was a large forum over five hundred feet long and a hundred feet wide, in which were found numerous statues, elegant columns and a triumphal arch.

Pompeii had been both sea port and seaside resort, a holiday haunt for all kinds of people – the emperor Claudius went there more than once. On some of the walls he might have seen the odd graffiti scrawl, an advertisement or an election notice. These various inscriptions show that the people not only spoke Latin but also Greek and a local dialect called Oscan.

There was a theatre seating fifteen hundred people and an amphitheatre which could accommodate at least twenty thousand spectators.

At one place the diggers found the charred remains of actual loaves of bread. These were circular and divided into triangular segments. Some loaves were found with the baker's name formed by the mould which had contained the wet dough.

But perhaps the saddest discoveries were the impressions of corpses. The dead bodies had long since rotted away, leaving hollows in the ash. Excavators pumped in liquid plaster and were able to recover the actual shapes of the once living people when the plaster hardened. One of the bodies was that of a dog with a collar.

Section 5 *Daily life*

Food and drink

One of the phrases often quoted when ancient Rome is being discussed is 'bread and circuses'. This referred to the custom, which had grown over the years, of giving poor and unemployed people food and entertainment to stop them rioting.

The 'annona', as the free food was called, started just over a century before Christ. There had always been shortages and local famines but by about 123 B.C. things had got so serious that poor people were allowed to buy cheap grain from government stores.

Fifty years later, the grain was free and was being supplied to 40,000 families. In another twenty years, the number 'on the dole' had gone up to 50,000 and during the reign of Augustus, almost a third of Rome's population (some 300,000 people) were being fed by the government. By that time, the grain was ground into flour and made up into loaves before being distributed.

Later still, there was free oil, pork fat and even wine. The annona lasted until the central government could no longer afford the huge expense.

Poor people ate bread when they could get it but preferred to boil loose grain into a kind of porridge, which was eked out with goats' cheese, eggs, olives, radishes or onions. Sometimes there was dried or salted fish. Meat was very rare, some poor families only eating it when it was given away after an animal had been sacrificed at a religious ceremony. So, no meat for the poor but a porridge of wheat, barley or millet for the afternoon meal at the end of the working day.

Most Roman tenements had no facilities for cooking beyond a little boiling on a portable brazier. If poor families wanted anything better, they either got the baker to cook for them or bought something ready prepared at a nearby pie or sausage shop.

A main meal at 1.30 or 2 p.m. was so early, that it was often necessary to have a light snack before bedtime. The poor man's breakfast wasn't very different from that of the well-to-do. Many people didn't eat breakfast at all, whether they were rich or not: others had a stale roll dipped in watered wine.

The main meal for the better-off Roman started a little later in the day – perhaps at 2.30 or 3 p.m. This

A cooking stove

A Roman kitchen. The toilet was normally in the kitchen.

56

may have been because of the growing habit of spending much of the earlier afternoon at the baths. Whatever the cause, the chief meal for the rich began later and later as time went on – and it lasted for several hours.

The banquet was usually given by a man for eight guests. Women and children sometimes ate separately in another part of the house. The guests stretched out at full length on three couches, each holding three diners, around three sides of a table in the 'triclinium', or dining-room.

They supported the upper half of the body on the left elbow, leaving the right hand free to put the food in the mouth. Forks were unknown and even knives and spoons were fairly rare.

A rich household had many slaves to prepare and serve the meal but it was quite common for a guest to take his own slaves with him to a dinner party.

It's probable that some of the stories of luxurious banquets that have come down to us are exaggerations. However, we know that meals could last for hours and consisted of many dishes to each course.

Shellfish, eggs, edible snails and some vegetables might be offered as a first course. The second could be sea fish, small roasted birds such as thrushes, cuts of wild boar, larger birds – not only ducks and chickens but also game birds (pheasants, partridges, et cetera), plus exotic species – cranes, parrots, flamingos or ostriches.

The main course had just as much variety – for example, venison, hare, sow's udders, sucking pig, ham boiled with figs and bay leaves, covered in pastry and baked with honey, plus turbot, salmon or sturgeon and all seasoned with pepper and fish stock.

The diners finished with honey cakes and fruit – not only that grown in Italy but rarer kinds from abroad. The list includes apples, pears, olives, grapes, figs, pomegranates, peaches, apricots, plums, quinces, mulberries, cherries, rhubarb and dates.

Each course was accompanied by wine. Although cider and beer were drunk in parts of the empire, wine was the rule in Rome. It was Italian grown and also imported from North Africa, Spain, Portugal, Germany and even from Palestine, Syria and Babylon.

Dining in a triclinium

Section 5 *Daily life*

Clothes

From earliest times, the patricians, or rich people of Rome, had settled on a clothing style that continued with little change for a thousand years. Everyone wore a tunic – men, women and children. Plebeians, or poor people, mostly made do with the tunic alone, whilst the well-to-do draped a huge swathe of cloth in folds around the body. This was called a toga.

Togas became more or less compulsory for all free born citizens whenever they went outside the house. The toga was at first always made of wool – and as it was shaped like the edge of a circle over two yards wide and six yards along the straight side, it got very warm in the summer.

After many years of the empire period, it slowly became permissible to leave the toga at home when venturing out. By a very late date, the only use a Roman had for a toga was to cover his corpse before his funeral.

The tunic for men and boys was a simple tube of material with openings for the head and arms. Its lower hem came to about the knees. In cold weather you could wear more than one. The emperor Augustus, who didn't like the short but nippy Roman winters, customarily put on no fewer than four tunics!

In early times, you rose in the morning, having slept in your tunic, with perhaps a toga or cloak as bed-clothes. You tied a leather thong round your middle and held the toga's end in one hand, while you arranged the folds round and round your body and over one shoulder.

Romans rarely had much of a wash in the mornings, preferring to wait for the visit to the baths in the early afternoon when work was over. Because the toga always had to look clean, Romans had to have at least one change of clothing.

It wasn't always obvious that garments were grubby for they were made of naturally-coloured sheep's wool. Your slaves could bleach them by laying them out in the sun but there were times when a man wanted a dazzlingly white outfit – perhaps when he was hoping to be elected to a public office. In that case, his things were whitened with wet pipe clay. Our word 'candidate' comes from 'candidus', a Latin word for 'white'.

Working man in tunic

Man in tunic and cloak

Wealthy man in tunic and toga

Some colours were allowed in order to give a little variety. For example, a bride traditionally wore yellow and reddish orange. Young boys sported a thin purple band along the edge of the toga but the only adults allowed the same privilege were senators and magistrates. The emperor alone was entitled to a completely purple outfit.

Women could have their ankle-length tunics and overmantles coloured or patterned. Underwear was not all that popular, although both males and females sometimes wore a kind of loincloth or pants. Mosaics have come to light showing lady athletes wearing bikinis with both bra and pants. In London, archaeologists dug up what looks like a small boy's brief leather pants, fastened at the hips with thongs.

Eventually, cotton, linen and silk were brought to Rome by traders. Linen comes from flax which Romans found they could grow in Italy itself but cotton and silk had to be imported from India and China. The new textiles were rarely used, however, for they always cost more than wool. Indeed, silk, for instance, could only have been afforded by the immensely wealthy – an emperor, perhaps – because in modern terms it must have cost several hundred pounds an ounce.

The founding fathers insisted that their womenfolk spun and wove woollen thread by hand. This continued for centuries in some old-fashioned families. It is well known that hand made things are expensive but it may be surprising to find that they didn't wear too well. This was almost certainly because clothes were washed by being spread on a rock at the riverside and pounded with a lump of stone!

A fuller, or laundryman, could be hired to soak your washing in urine and to jump up and down on it with his bare feet, or to treat it with fuller's earth.

Sandals could be thick-soled and hobnailed for soldiers, workmen, or for ordinary country wear. Shoes of lighter, flexible leather were worn in towns. In rural areas, hats were common but in Rome, a fold of the cloak or toga was held over the head when it rained.

The womenfolk of the first Roman settlers parted their hair in the middle and dragged it back severely. Later on, all sorts of fancy hair styles were permitted. Men started by wearing long hair and beards. They were clean shaven towards the end of the republic and grew beards again in the second century, from the example of Hadrian. Women wore large pieces of gold jewellery and painted their faces. Unfortunately, these paints were often metal oxides which were bad for the skin.

A typical man's hairstyle

Woman in tunic

Woman in tunic and cloak

Women's hairstyles ranged from simple and austere to very elaborate

59

Section 5 *Daily life*

Education

A school room

It's natural to suppose that the children of the first citizens were taught merely by watching their parents and copying their actions and methods. Roman parents continued to educate their own children personally and it wasn't until later that families employed Greek slaves as private tutors or sent their children to the local school.

The elements of reading, writing and arithmetic were taught at the primary stage by a 'magister', or teacher. He was his own boss, running his 'school' in the curtained-off corner of a shop or a disused room of someone's villa. He charged each pupil a few coppers a week and as there weren't usually more than ten to twenty young students, he can't have been very well off.

Lessons consisted of learning things by heart and copying out huge tracts from Latin authors. Rote learning of this kind also included tables of measurement – for instance, length, liquid measure, area and money.

Pupils had to know that there were 'ten asses in a silver denarius'. The 'as' was also a unit of weight. This was a twelve ounce pound, but the word 'as' came to be applied in later years to a copper coin weighing a mere half ounce.

The tables were extremely complicated and a failure to chant the right names and amounts earnt a stroke of the cane.

The copying was done on a writing tablet with a stylus. The tablet was a pair of wooden boards with raised surrounds and hinged together with a leather thong. The flat parts were covered with wax and the letters were inscribed with the sharp end of a stylus, or pen. Mistakes could be removed by smoothing over the errors with the flat end of the stylus.

Roman arithmetic must have been particularly hard to do, since numbers were represented by letters. It isn't impossible to do multiplication and division with these letters but it's very difficult. The chances are that Romans counted on their fingers or used a sort of abacus with pebbles fitting into rows. (Our word 'calculate' comes from the Latin word for a pebble.)

60

School started at sunrise and went on with no break until lunch time in the middle of the day. The school year began in the autumn and continued until the following high summer when the magister closed up for the annual holiday.

Boys, when they reached the age of about twelve, went on to another kind of teacher called a grammarian. Girls of that age, if they had attended the magister's classes, then went home and had very little further education.

The grammarian concentrated on the Greek language, partly because of Rome's everyday business and administration dealing with the eastern Mediterranean, where nearly everyone spoke Greek. The other main subject was the art of public speaking. No rich man's son could hope to be elected to any important position in the government without a thorough knowledge of how to deal with people and what to say if you wanted to change their opinions.

To round off his education, a wealthy young man might start by taking clients and pleading their cases in law courts. No examination had to be passed. The only requirement was his ability to defend his client properly.

Another route leading to well-paid jobs was to be appointed as an officer in the Roman legions. This often led to men becoming governors or magistrates.

Such was the education of a typical Roman youth of good family.

Examples of writing

Boys writing on waxed tablets with styluses

Section 5 *Daily life*

Time and the calendar

A stage direction in Shakespeare's play, 'Julius Caesar', asks for a clock to chime. Unfortunately, these were unknown in Caesar's time, so our great dramatist seems to have made a mistake.

How did they tell the time in Rome? There were, of course, twenty-four hours in their day as there are in ours. The difference was that the Romans then divided the day into twelve hours and did the same for the night. This meant that an hour of a summer's day was a lot longer than one in the winter, for the daylight lasts several more modern hours in June than it does in December.

Romans probably made appointments by saying, 'I'll meet you at the fifth hour', knowing that the time chosen was just less than half-way through the day. You couldn't expect someone to call at your house at exactly two o'clock, if no one knew precisely what the time was.

Romans did have water clocks, similar to the ones used in Egypt. One version of the 'clepsydra', as these clocks were called, had a glass cylinder full of water, which ran out through a tiny hole in the bottom. There were marks scratched onto the outside of the glass (or on the inside, if it was a metal cylinder) so you could see how far the water had sunk down and thus, the hour.

A sundial

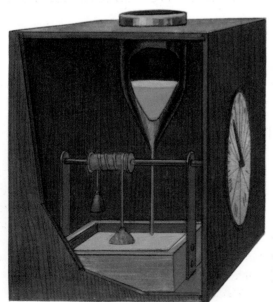

A water clock. In this version the hand on the clock turns when the water level rises

Sundials were first brought to Rome from Greek cities but such timepieces have to be made for the place where they are to work and the Romans took time to get their sundials to work properly in Italy.

Nowadays we know that the earth doesn't go round the sun in an exact number of days. The Romans didn't know – nor were they aware of the precise number of whole days involved.

They based their calendar on lunar months (that is, the time from one new moon to the next), but twelve lunar months only made 355 days. The result was a calendar getting farther and farther away from the season it was supposed to tell. If you believed there were only 355 days in the year, you would be starting your second year when there were still ten days to go. In the second year you'd be the original ten and a bit days out, plus another stretch of ten and a bit days. In five years, the calendar would be out of step with the seasons by more than seven weeks and another twelve years later, you'd find yourself celebrating midsummer day in the middle of the winter.

The Roman calendar had twelve months – four named after gods and goddesses and the others known

DAYS OF OUR MONTH	JANUARY AUGUST DECEMBER				FEBRUARY				APRIL JUNE SEPTEMBER NOVEMBER				*MARCH MAY JULY OCTOBER			
1	Calends				Calends				Calends				Calends			
2	a(nte)	d(iem)	IV	Nones	a.	d.	IV	Nones	a.	d.	IV	Nones	a.	d.	VI	Nones
3	a(nte)	d(iem)	III	Nones	a.	d.	III	Nones	a.	d.	III	Nones	a.	d.	V	Nones
4	Pridie			Nones	Pridie			Nones	Pridie			Nones	a.	d.	IV	Nones
5	Nones				Nones				Nones				a.	d.	III	Nones
6	a.	d.	VIII	Ides	a.	d.	VIII	Ides	a.	d.	VIII	Ides	Pridie			
7	a.	d.	VII	Ides	a.	d.	VII	Ides	a.	d.	VII	Ides	Nones			
8	a.	d.	VI	Ides	a.	d.	VI	Ides	a.	d.	VI	Ides	a.	d.	VIII	Ides
9	a.	d.	V	Ides	a.	d.	V	Ides	a.	d.	V	Ides	a.	d.	VII	Ides
10	a.	d.	IV	Ides	a.	d.	IV	Ides	a.	d.	IV	Ides	a.	d.	VI	Ides
11	a.	d.	III	Ides	a.	d.	III	Ides	a.	d.	III	Ides	a.	d.	V	Ides
12	Pridie			Ides	Pridie			Ides	Pridie			Ides	a.	d.	IV	Ides
13	Ides				Ides				Ides				a.	d.	III	Ides
14	a.	d.	XIX	Calends	a.	d.	XVI	Calends	a.	d	XVIII	Calends	Pridie			Ides
15	a.	d.	XVIII	Calends	a.	d.	XV	Calends	a.	d.	XVII	Calends	Ides			
16	a.	d.	XVII	Calends	a.	d.	XIV	Calends	a.	d.	XVI	Calends	a.	d.	XVII	Calends
17	a.	d.	XVI	Calends	a.	d.	XIII	Calends	a.	d.	XV	Calends	a.	d.	XVI	Calends
18	a.	d.	XV	Calends	a.	d.	XII	Calends	a.	d.	XIV	Calends	a.	d.	XV	Calends
19	a.	d.	XIV	Calends	a.	d.	XI	Calends	a.	d.	XIII	Calends	a.	d.	XIV	Calends
20	a.	d.	XIII	Calends	a.	d.	X	Calends	a.	d.	XII	Calends	a.	d.	XIII	Calends
21	a.	d.	XII	Calends	a.	d.	IX	Calends	a.	d.	XI	Calends	a.	d.	XII	Calends
22	a.	d.	XI	Calends	a.	d.	VIII	Calends	a.	d.	X	Calends	a.	d.	XI	Calends
23	a.	d.	X	Calends	a.	d.	VII	Calends	a.	d.	IX	Calends	a.	d.	X	Calends
24	a.	d.	IX	Calends	a.	d.	VI	Calends	a.	d.	VIII	Calends	a.	d.	IX	Calends
25	a.	d.	VIII	Calends	a.	d.	V	Calends	a.	d.	VII	Calends	a.	d.	VIII	Calends
26	a.	d.	VII	Calends	a.	d.	IV	Calends	a.	d.	VI	Calends	a.	d.	VII	Calends
27	a.	d.	VI	Calends	a.	d.	III	Calends	a.	d.	V	Calends	a.	d.	VI	Calends
28	a.	d.	V	Calends	Pridie			Calends	a.	d.	IV	Calends	a.	d.	V	Calends
29	a.	d.	IV	Calends					a.	d.	III	Calends	a.	d.	IV	Calends
30	a.	d.	III	Calends					Pridie			Calends	a.	d.	III	Calends
31	Pridie			Calends									Pridie			Calends

*In March, May, July and October the days of the Nones and the Ides were set respectively on the 7th and 15th, instead of the 5th and the 13th as in the other months.

'Pridie' means 'day before'.

The Roman calendar after Julius Caesar

as 'the eighth month', 'the fifth month', or whatever it was The calendar seems to have run originally from March to February, hence December was the tenth month. There were additional months put in from time to time to keep the calendar straight. The first of each month was called the 'calends', the mid-month day the 'ides' and the first fortnight thus formed was split in two by days named 'nones'.

Eventually, it was realised that something would have to be done to make the calendar more accurate. It was left to Julius Caesar to make the alterations. These were brought about in the year 46 B.C.

Later still, 'Quinctilis' and 'Sextilis' ('fifth' and 'sixth' months) were renamed in honour of Julius and Augu-stus Caesar (July and August). We still refer to the original seventh month as 'September' and the rest of the autumn and winter months by their Latin numbers. Compare 'October' (eighth month) with 'octopus' (eight legs) and 'December' (tenth month) with 'decimals' (tenths).

Julius's new calendar involved the addition of ninety extra days to 46 B.C. in order to get the calendar in time with the solar year, but his rearranged system lasted until about two hundred years ago, when it had to be reorganised once more. At that time, the year had to *lose* eleven days and groups of foolish people went about in bands, shouting, 'Give us back our eleven days!' – just as if they had really been robbed of something.

Section 5 *Daily life*

Painting, sculpture, drama and literature

If you've ever visited the remains of a Roman building, you'll have almost certainly seen at least one example of Roman art, namely, a stone-patterned floor. Often, these mosaics are merely geometric designs, made by arranging thousands of little squares of differently coloured stones. Sometimes there are actual pictures – a human head, flowers and leaves or an animal.

If you want to see proper Roman painting, you'll probably have to go abroad. There are some small fragments of coloured wall plaster in museums, such as the one at St Albans, but the most complete pictures are to be found, for example, at Pompeii in southern Italy.

A Roman theatre in Orange, southern France

Romans occasionally painted on panels of wood but the commonest specimens come from the interior walls of buildings. Wallpaper is a fairly modern invention, so Roman householders had to choose between bare plaster for their rooms, or (if they could afford it), hand-painted illustrations of legends, their own history, sporting activities, landscapes, or perhaps still-life pictures of food and elegant tableware.

Most well-to-do Roman homes had a statue or two in the garden. These might have been made by a local artist as original works but were more likely to be copies of Greek figures, turned out by the dozen in the studio of a man who was more businessman than sculptor.

For their public buildings Romans liked to have the best models that could be had: in most cases, this meant importing them from foreign countries, now part of the empire. Roman generals plundered defeated Greece and sent thousands of priceless statues back home to grace the walls of the latest temple or bath house.

But it wasn't all a case of Greek figures or nothing. The Greek examples were (and are) very beautiful, no doubt, but local Roman artists could produce a stone head with all its character and oddities so lifelike that the friends of the sitter could recognise it at once.

The examples of statues and busts that have come down to us include portraits on tombs and decorations on triumphal arches and other monuments. Among the latter is Trajan's column in Rome which has some 2,500 figures sculpted in a kind of strip cartoon winding round the pillar from bottom to top in a 215 yard long account of the emperor's campaign in Dacia.

For entertainment, a Roman citizen could go to the theatre but rarely with the chance of seeing the high standard of play the Greeks had enjoyed. The building where the plays were shown, unlike its counterpart in Athens, was completely enclosed within high walls. There was also an arrangement of pulleys and hooks so that the audience could be sheltered from the fierce summer sun by a huge canvas awning which was drawn high over the seats.

Plautus wrote over a hundred plays – mostly musical farces. Terence also wrote comedies. There were also writers of tragedy such as Livius Andronicus, Naevius and Seneca. Unfortunately, the normal standard of drama was never quite as high as that of Greece, most Romans preferring to see low pantomime rather than Greek originals.

For reading matter, citizens had to rely on hand-copied books, for printing had yet to be discovered. Books were in the form of rolls and were usually kept in a container like an umbrella stand.

In those days, a writer could be an expert on many different subjects, although authors such as Cato, Livy, Caesar, Tacitus and Suetonius are known mainly for their history writing. Sometimes, Roman history was not quite as scientific as modern historians would like.

Other authors include poets – Ovid, Horace, Catullus and Virgil. Virgil's poetry could be about almost anything – bees, cattle-breeding, quality of farm soil, weather signs, trees, vine-growing and many other topics. His chief claim to fame, however, was the *Aeneid* which gave Romans an outline of this largely mythical 'history'. Cicero was famous for his essays and speeches.

Although Roman literature, like much of their art, began by copying Greek originals, it eventually achieved a life and status of its own.

Section 5 *Daily life*

Earning a living

In some prehistoric villages, there were so few different kinds of life-style that it is possible to describe such a village in a couple of pages. All you have to do is to talk about a typical villager, knowing that his daily experience will be more or less the same for everybody. Rome was no village, though – it was the capital of a great empire, with more than a million inhabitants by the time Christ was born. Life was more complicated and there were thousands of different kinds of jobs to be done. We can't cover all the possible ways a person could earn a living – all we can do is to pick some examples.

If you were rich, you probably owned at least one estate (maybe more), possibly farmed by slaves. You might even employ a superior slave to run your properties for you. From this class of estate owners in earlier times came all the senators and other senior officials, both in the army and outside it.

The next class were called 'knights', as they had once been at least wealthy enough to ride a horse into battle. The Latin word for them was 'equites' from 'equus' (a horse). These were the men who either owned (or could raise) enough money to finance public building contracts, or to fit out ships for overseas trade.

Neither of these activities was allowed to senators. Knights tended to keep out of politics, so that they could run the empire's trade and businesses and make fortunes for themselves. They might be found superintending the building of a temple or an aqueduct: they might be running a spinning and weaving workshop.

To help them in the clothing trade, they employed spinners, weavers and fullers. The last mentioned, you may remember, were also the laundrymen of Rome, using fuller's earth to absorb grease from oily fleeces as well as from dirty clothing.

You might find a knight in charge of a mass production pottery, a mosaic maker's workplace or perhaps a jeweller's shop. He wouldn't, of course, do the work himself but employ others to carry out his orders.

Lower down in the social scale were the owners of small shops. There were no chain stores in the old city and each shop was owned and run by a separate family. Some streets were devoted to one type of shop only – for example, all the harness makers lived and worked within sight of each other as did the glass makers.

As we've seen, the owner of the shop was normally a craftsman who made the product in a workshop at the rear of the premises and sold it from a counter on or near the street. In Rome you could see the shops of shoemakers, tailors, butchers, greengrocers, drapers, carpenters, wine sellers, barbers, herbalists, tanners, booksellers, rope makers, paint sellers, perfumers and many more.

Not all of the above were makers and sellers – some just sold, others provided a service. You could go to a joiner's to buy a table or a cupboard, or you could hire a carpenter to do woodwork in your own house. Other providers of services included fortune tellers, doctors and teachers. There were no specialised lawyers – anyone could speak on behalf of the accused in a law court.

There were also civil servants such as secretaries (often slaves), fire wardens and building inspectors.

More specialised jobs were done by armour makers, horse breakers, trappers of wild animals and owners of gladiator schools (at least in the early days) where fighters were trained for the Colosseum and other amphitheatres.

Some men did their jobs away from the city centre – brick and tile makers, stone cutters, most builders, engineers, shipwrights, sailors and soldiers. If you had no other skill you could always become a porter or delivery man. No carts were allowed through the city gates during the hours of daylight, so the goods they carried had to be unloaded and shouldered on their way by armies of carriers.

If you wanted to make a fortune and didn't mind an element of danger you could always become a paid gladiator or even a chariot driver in the races at the circus.

A shop selling bronzeware, including lanterns, jugs, and lamps

Section 5 *Daily life*

Games and pastimes

The people who lived in Rome spent more time out-doors than we do. This was partly to do with climate. Even today, you won't want to stay in your house when the temperature is high and the weather good.

In those days there wasn't a lot to stay in for: there were no radios or televisions and the light at night in a Roman flat was scarcely bright enough for reading. As a result, Romans got up when the sun rose and many of them went to bed when it set.

It's obvious that most of their pastimes were open air ones. The extremely popular animal and gladiator shows, together with chariot-racing are dealt with in Section 7, so little will be said about them here. One point of interest is that the vast majority of Roman men were extremely keen on betting. They would gamble on the result of a fight in the arena or on the outcome of a race in the circus.

Betting was reckoned by the authorities to be such a social sin that it was strictly forbidden – except in the arena or at the circus. It was considered all right to bet on a gladiator or a chariot driver but not on anything else. Not that this worried the average citizen, however. He knew perfectly well that there were a dozen wine shops or taverns almost within sight of his own home, each of which had an illegal back-room betting shop.

The gambling was on the roll of a pair of dice. There are a number of ways of doing this; the simplest being to throw a higher number than your opponent.

Knuckle bones and dice were used in gambling

Men playing dice

68

At Saturnalia (about the same time as our Christmas), betting was allowed on anything – even coin games. Someone would spin a couple of coins and men would bet on whether they came down two heads, two tails or one of each.

Another popular 'game' was played by two people facing each other. Both hid their right hands behind their backs and at a signal both showed their opponent the previously hidden fist, this time with one to five fingers extended. At the same time, each player called out a number from two to ten. This was supposed to be a guess at the total number of fingers showing. The winner was the one who guessed correctly.

As well as 'micatio', as this game was called, there were more serious 'sitting down' games such as the Roman versions of backgammon, chess or draughts. Board designs, similar to the ones on which we play draughts, have been found scratched into the hard stone pavements of the cloistered walks around Rome's forums.

More energetic were the ball games or wrestling a Roman might prefer when he went to the baths. Some rich citizens enjoyed yachting and those less well off could take part in angling contests. If all you wanted was peace and quiet, a gentle stroll around the public squares after the law courts had finished for the day might be your choice, or even a walk through the grounds of the emperor's palace.

Some popular children's games from Roman times are shown in the pictures on this page.

Miniature chariot

Boys playing 'the mill game' – a cross between noughts and crosses and draughts

Boy playing with a hoop

Government and the law

This is a reconstruction of the centre of imperial Rome in the mid second century, showing some of the official buildings and temples of the central government.

We may remember how the ruling of Rome began in the early days with the choosing of a monarch – until the very idea of kings became unacceptable. When Tarquinius Superbus had been driven out, two consuls were appointed to lead the citizens for one year at a time, both of whom had to agree before any law was made or scrapped and before war was declared or peace sought.

In time of war, a military dictator was appointed to lead Rome's soldiers against the enemy. He was ex-pected to give up his powers as soon as the emergency was over.

A group of (mainly) ex-consuls formed the city's first parliament, or senate, as they called it. Senators were elected by the landowners and there were usually about three hundred of them. Poor people had their own assembly but it had little power. The two classes were known as patricians (the rich) and plebeians (the poor). A great deal of Rome's history was the running battle between the patricians and plebeians – the former trying to hang on to all of their power and the latter aiming to prise some of it away.

The plebeians chose their moments well: they waited

Looking down on the Roman forum, from the Temple of Jupiter on the Capitol, about 150 A.D. The building bottom right is the Basilica Julia where trials and business matters were conducted.

Seventy years later, Claudius set educated ex-slaves to run various government departments and the senate was much less powerful than it had once been. The emperor Domitian was so contemptuous of the senate that when some of its members questioned his actions, he had them executed!

Hadrian had to order some patrician rebels to be killed and then, because the senate complained so strongly, he had to promise never again to do so without the senate's consent. Antoninus Pius, on the other hand, always consulted the senate before doing anything important – such as spending public money!

The emperor Pertinax was murdered and the army put the empire up to the highest bidder at an auction. The senate refused to accept the buyer, Didius Julianus, as their emperor. There was yet another civil war; Septimius Severus became the next ruler and Julianus was executed. Septimius had realised that the army was the body which held the gift of power, not the senate. He therefore raised army pay and took away very nearly all of the senate's last privileges.

From then until the final collapse, the emperor's word was law, provided that he could afford to bribe the troops and stay in power long enough to have his wishes respected. Apart from that, the army was the law.

Oddly enough, it was Roman law which survived the empire. To begin with, the law had concerned itself with quarrels between people. Justice was a very rough and ready affair. As time went by, the government took a larger and larger interest in crime. Also, in earlier times, the law's protection in the matter of wills, inheritances, divorce, private disputes and so on was reserved for the upper classes: if you weren't a 'citizen', you couldn't claim anything legally.

For the poorer early settlers, the law had seemed unjust – none of it was written down and only the upper classes had the right to say what it was. After protests, the laws were published – they were engraved on publicly displayed bronze plates. Unfortunately, these disappeared during the Gaulish invasions, so we only know of them from references in ancient writings.

No exams were needed to be a lawyer – only a persuasive tongue. Many young men began their careers by speaking in the law courts.

In spite of the defendant's habit of presenting the judge with a gift (a custom that sounds like bribery!) the Latin system prospered and grew. It was taken all round the Mediterranean and many modern codes of law are based on it. Our own lawyers use Latin words in their everyday work – words such as judge, jury, verdict and justice.

until Rome was in danger from an enemy attack and then went on strike against army service until the patricians allowed them to elect a tribune to look after their interests.

Towards the end of the republic, those patricians entrusted with high military command tended to forget about the rule asking them to step down when the crisis was over. Then violence was often the outcome. Julius Caesar was murdered because it was thought that he might want to be king.

Eventually one man – Augustus – succeeded in staying at the top, whilst pretending that there was still a republic. But he was, in fact, the first emperor.

Religion and legends

Early Roman religion was based on a belief in spirits that were invisible. These spirits were called 'numina' and there was a 'numen' for practically everything – a numen for the night, the day, each hill, home, river, mountain, lake, forest, field and almost anything with a separate identity.

Before taking any important step it was necessary to plead with the proper numen, whose shrine might be on a hilltop, in a grove, at a crossroad or in a cavern. A farmer would plead for the numen's help before planting or reaping a crop and a merchant might do the same before taking on a contract to supply bricks.

It was essential to get the words and actions right and to know what offerings to bring the numen. The men who could remember all these things became the first priests. They weren't the spiritual leaders we expect our clergymen to be – they merely knew who to bribe for heavenly help and how to do it.

It may have been the Etruscans who taught Rome that it was easier to think of a deity if a statue was made and set in a specially built house or temple. It was almost certainly they who introduced the gods and goddesses of Greece to the citizens of the seven hills. Many of these were simply taken over, merely changing the name to a Roman one.

In this way, the character of Zeus was adopted as Jupiter and Hera taken over as Juno. The Greek Artemis became Diana and Athena was renamed Minerva. From Ares, the Romans had Mars and Hermes changed into Mercury. Venus started in Greece as Aphrodite.

The most important god of later times was Jupiter, god of light and king of the immortals. The specialities of some of the others were as follows: Juno (queen of the gods), Saturn (agriculture), Ceres (crops), Minerva (wisdom), Venus (love), Vulcan (fire), Neptune (sea), Diana (moon), Apollo (sun), Mars (war), Vesta (hearth), Manes (the dead) and Janus (doorway).

As well as these, there were personal and family gods – the Lares looked after the home and the Penates the store cupboard. The father of every Roman family set up, where possible, a shrine to these protectors and led whatever prayers were said.

Legionaries returning from overseas conquests

Jupiter Optimus Maximus – 'the best and greatest'. There was a temple to him, Juno and Minerva in every Roman town.

brought back the worship of foreign gods such as Mithras from Persia and Isis from Egypt. In later times, as belief in the classical gods dwindled, some Romans took up the study of new philosophies, such as that of the Stoics who were encouraged to ignore both pleas-

Making an offering to the Lares and Penates – the household gods

ure and pain, or of the Epicureans who believed in the pursuit of pleasure as the greatest thing in life.

The custom arose in the days of the empire of making the last dead emperor a god, or even the supreme god. The worship of the departed ruler, as with that of the other gods, was regulated by the government who appointed the priests.

The high priest was known as 'Pontifex Maximus' and the honour normally went to the living emperor. There were other kinds of priests, for example, the flamines who were the burners of offerings and the augurs who could tell the future from flashes of lightning, the flight of birds, or the livers of sacrificed animals.

Romans didn't have regular church meetings with prayers; their visits to temples being more in the nature of bargains struck with the god – 'You protect me on my journey and I'll give you an offering of food and wine.'

In the matter of legends, the Romans were content on the whole to retell Greek myths. Their own history provided some examples of folk tales. These were usually designed to show how brave or noble their ancestors were. We've heard the stories of Romulus and Remus, the adventures of Aeneas, the legend of the Sabine women, to mention only three. Other legendary figures were Cincinnatus, who was made military dictator during a crisis. The period of his authority was supposed to run for six months but he left his plough, beat the enemy in a few days and went straight back to his farm.

In the Punic Wars, a Roman general named Regulus was captured by the Carthaginians. They released him on his honour to return at the end of his mission. This was to go to Rome and persuade his countrymen to surrender. On the contrary, he urged them to go on with the war. Then, because of his promise, he returned to Carthage, even though he knew he was going to torture and death.

Another hero was the magistrate whose sense of duty was so strong, he tried his own sons for rebellion and sentenced them to death.

That was the sort of person the average Roman liked to think he was – brave, stern and dutiful.

The Temple of Vesta, showing the sacred fire

Christianity

The life and death of Jesus passed almost unnoticed in the Roman world. No Roman writer of the time thought it important enough to mention. If they had heard of him at all, they probably dismissed him as another unsuccessful eastern revolutionary.

The followers of Jesus claimed to have talked with their leader after his crucifixion. He had told them to spread the word of God and to tell everyone about the forgiveness of sins and the life everlasting.

His disciples, or followers, started to carry out his request and preached mainly to the groups of Jews scattered throughout the empire. On the whole, however, it wasn't the Jews who were converted but those whose lives were so poor and downtrodden that they embraced Christianity willingly.

The Romans thought (mistakenly) that these converts were all Jews – after all their leader had been one, hadn't he? Not that it mattered much: Rome had always welcomed new religions from all over the civilised world – wasn't this just another new faith? In any case, no Roman thought it odd to believe in more than one god: if a prayer to Jupiter was good then prayers to a couple of other deities were even better.

It wasn't long, though, before the differences became extremely clear. Christians were not like other worshippers – they wouldn't accept office from the government, they wouldn't serve in the army and they refused to bow down to statues of the emperor-god. Neither would they make offerings of food and wine – either to him or to any other of the accepted host of Roman heavenly beings.

St Paul (like St Peter) came to preach in Rome itself. He is supposed to have made converts among the official staff of the emperor. At one stage it was ordered that every citizen must make a sacrifice to one of the accepted gods in the presence of a magistrate. He would then be given a certificate. This, the authorities thought, would make it easier to detect Christians who

Catacombs

Many Christians were killed by lions in the arena. This was a normal method of execution

came before them.

The emperor Nero was suspected (with some reason) of having started the great fire of Rome. Christians were convenient scapegoats and he had them arrested by the hundred. Many were thrown to the lions in the arena: some were burnt to death.

This was only the first of many waves of persecution – strong feelings were whipped up against them, arrests made and executions carried out. Then things would quieten down until the next emperor who felt it was time to make the Christian community toe the official Roman line. Both Peter and Paul perished during the persecutions.

To escape their accusers, Christians often took refuge in the miles of tunnels that ran beneath the city. They were called 'catacombs' and were used to bury the dead. Here the refugees hid, arranged services and held meetings.

Followers of Christ also developed secret signs so they could be sure of revealing their beliefs to friendly ears. A man might trace the outline of a fish with his finger or the end of his staff. This was because the Greek word for 'fish' was made up of the initial letters of their leader's name and title (Jesus Christ, Son of God and Saviour).

As time went by, the average Roman believed less and less in the official gods. More and more citizens became Christians – not just the poor and lowly but people from every class of society. Even a few of the patricians were claimed for Christ. Many of the latter had tired of an endless round of pleasure and were attracted to what the new religion had to offer.

By the early fourth century, belief in Christ had become respectable and in the reign of Constantine, Christianity was elevated to an official Roman religion. Constantine is reported to have seen a vision of a cross during a battle and to have become converted.

Christian churches were built throughout the Roman world and when the empire eventually broke up, it was the Church which preserved the best of the ancient world and passed it on to the future.

Slavery

Prisoners of war – men, women and children were sold to slave-dealers

'Did you want to speak to me?'

'Yes. We'd like to know about slavery. I believe you are a slave: is that correct?'

'I'm a slave all right. My name is Menenius Lucius Centullus. It's not my real name but one given to me by my owner. I can't remember my proper name but I know my family came from Britain during Julius Caesar's raids.

'I suppose my ancestors must have cost a lot in those days. You see, most slaves are actually captured enemy soldiers or their descendants. If your parents are slaves, then so are you.'

'How were your ancestors bought and sold?'

'Why, in the slave market, of course. It was very undignified being paraded around like a prize sheep so that the public could bid for you. As I was saying, slaves were expensive in the early days but as Rome's empire grew, so did the numbers of prisoners of war. These were auctioned off by the dozen, the hundred and eventually by the hundred thousand.

'Naturally, there were few rich families which could

afford the early prices but as the numbers grew the cost fell rapidly. Nowadays, most citizens who have houses own at least half a dozen slaves. Rich people own hundreds and men like the emperor have thousands.

'The price of a slave also depends on what he or she can do. The better educated and those used to town life fetch the highest amounts and are taken by merchants to act as clerks, book-keepers, secretaries and such like – or they go to be house servants.

'Those with little learning but strong muscles may be sent to work in chain gangs on a farm or down a mine. If the person on sale is a soldier, he can be sent to one of the gladiator schools to be trained to fight to the death in the arena.

'In the old days, slaves were often branded or chained up to make escape difficult. Some slaves were actually kidnapped off the streets of Rome itself: it was very difficult to get away once that happened to you.'

'Were runaways punished?'

'Oh yes; a slave could be whipped, chained, starved or beaten to death – and not just for trying to escape either. If an owner wanted to ill-treat his slave in any of these ways or even to kill him, the owner wouldn't suffer. It was reckoned to be his right to do what he liked with his own "possessions".

'If a slave pretended to be a free man, he could be put to death. So he could if he tried to enlist in the army. Some were employed directly by the government on building or repairing things – temples, roads, aqueducts – that kind of job.

'I suppose I'm lucky. I work as an assistant in the library of the public baths. I'm probably better off than many free citizens. That does worry me a little bit.'

'How so?'

'Well, as time went on, public opinion turned against very harsh punishments. Did you know that the law actually says, "If a slave murder his master, not only he but all the other house slaves shall be executed"? That actually happened about two hundred years ago but I'm sure it couldn't happen now. Then again, I think the growth of Christianity had something to do with it. People are kinder. The trouble is that the kinder you are the more it costs to keep a slave. Then one day your master realises that it would be cheaper to set you free. I don't want to be free – all that would mean is unemployment.

'Once upon a time, a slave would try to save enough money to buy his freedom. The only other ways to escape were freedom under your master's will, rebellion or death. You've heard of Spartacus? He was a gladiator slave who escaped with some of his friends. They defeated the soldiers sent to kill them and took their arms and armour. They released other slaves until they were an army almost 10,000 strong. They terrorised southern Italy for two years. Then General Crassus beat them and rounded them up. He had them crucified – one to a wooden cross, every thirty yards from Rome to Capua, a hundred miles away.

'That's what being a slave was like in the old days!'

Slaves being crucified along the Appian Way after the revolt led by Spartacus

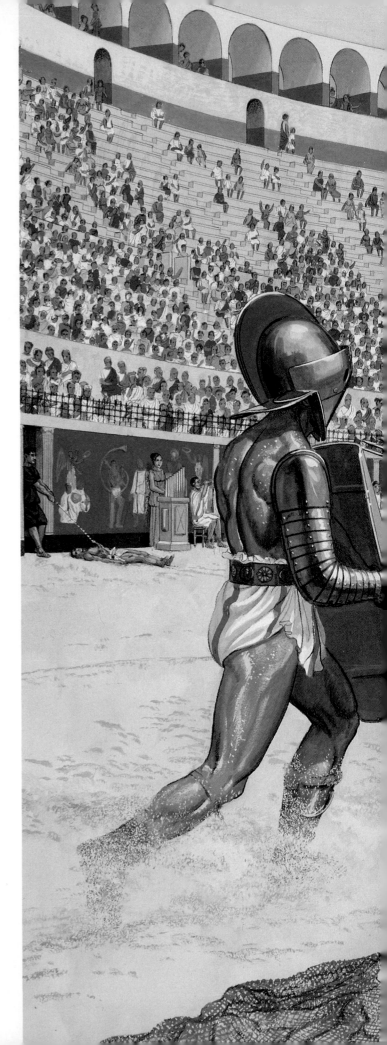

Section 7 *The arena*

The amphitheatre

Romans didn't care overmuch for the kind of play the Greeks supported but they used the Greek theatre ground plan in the construction of their amphitheatres.

There men and beasts fought one another for the amusement of the crowd. Every town of any size at all had an amphitheatre ('theatre on both sides') with rows of seats encircling and rising round a sandy space in the middle. This was called the 'arena' (the Latin word for 'sand'). You can find them all over the Roman Empire – from Portugal to Asia Minor and from Africa to north Britain.

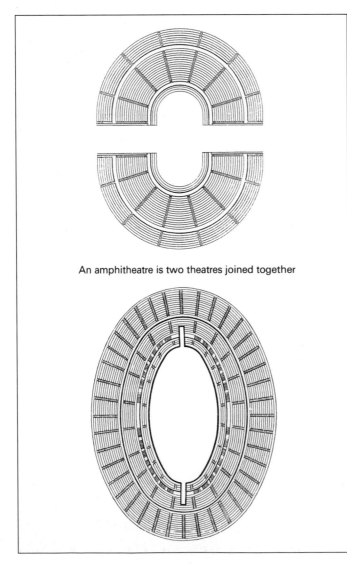

An amphitheatre is two theatres joined together

Gladiators fighting in the arena. A secutor fights a retiarius (see page 83)

The Colosseum

Whenever the subject of ancient Rome comes up in conversation, it is almost certain that someone will mention the inhuman Roman 'games'. These started in imitation of the customs of other Italian tribes. Perhaps the Samnites or Etruscans had the habit of holding elaborate funerals for their dead chiefs.

At first, the slain leader's slaves were gathered together and executed. Later on in history, they were given swords and paired off so that they could kill each other.

Romans took over this practice but used captured enemy soldiers. With a simple weapon each – perhaps a spear or dagger – they were encouraged to fight to the death, the winner's prize being his own life and freedom.

The fights were held in public squares, armed Roman soldiers forming a ring to prevent escapes. When Rome became the centre of an empire, the emperors began to put up special buildings to house the 'games', as they called them. The greatest of these was the Colosseum.

It wouldn't have been any good asking an ancient Roman to direct you there: the name 'Colosseum' wasn't used until centuries after the empire had passed away. The word comes from the 'colossal' statue of Nero which once stood on the site. When the Colosseum was new, citizens called it the Flavian amphitheatre, after Flavius Vespasian, who began it.

Vespasian had Nero's garden lake drained and the statue smashed to make room for the great building's foundations. It was near the centre of the city and was the greatest project of its kind in Rome. The emperor died before the work was complete and it was left to his son, Titus, to open it to the public.

Even then, it was still unfinished when the first performances took place, the last touches being added later. Finally, it was a gigantic stone oval, three to four storeys high (perhaps 150 feet), with outside measurements of 616 feet along its greatest diameter.

In spite of earthquakes and the pillaging of its fabric in the Middle Ages by those seeking easy and cheap quantities of building stone, enough remains to show modern tourists what might easily be added to any list of 'wonders of the ancient world'.

The Colosseum

The arena was formed of stone blocks covered with sand. The blocks rested on top of the walls of underground compartments – cells for the performers and cages for the animals. The oval arena proper was 262 feet by 177 feet and surrounded by a high metal fence to protect the crowd from the wild beasts.

Beyond the fence, there was a corridor right round the arena and then a smooth stone wall thirteen feet high. Above this began the tiers of seats – the first few rows reserved for the emperor and other important officials. Higher still sat the general public in what seem to have been numbered and reserved seats. The Colosseum could seat about 45,000 spectators with room for perhaps another 5,000 standing.

There were many corridors, staircases and exits so that the crowd could be dispersed quickly and safely after the 'show' was over. Many a modern football ground might envy the eighty or so controlled exits.

At the top of the building were the spars and pulleys controlling the canvas awnings which could be drawn across to shield the audience from the intense heat of midsummer.

At first, there was a system of water pipes which could flood the arena and turn it into a lake, on which mock sea fights between galleys could be fought. For the fighters, it was anything but pretence. The boarding parties had swords and spears and there were always soldiers surrounding the 'lake', ready to put an arrow into anyone seen trying to slip away by swimming to the side.

Section 7 *The arena*

Gladiators

It has to be said that many Romans (perhaps a majority of them) were cruel and bloodthirsty. They were like ignorant and brutal peasants who had suddenly become rich enough to give in to whatever beastly passion they liked. Nothing else can explain centuries of men being butchered for sheer amusement.

We know that some prisoners of war were trained as gladiators – the fighters to the death. A few criminals were also taken and there were those who volunteered because they were down on their luck and could earn money if they won.

Why did anyone consent to be a gladiator? If you were a learner and were told, 'Either you kill your opponent or we'll kill you', the chances were that you'd do what you had to, to stay alive.

The learners were kept in small stone cells in training schools and only allowed out to practise. To make sure nothing dangerous happened when they were training with real weapons and that there was no likelihood of escape, the exercise yard was surrounded by guards armed with bows and arrows.

The men rehearsed with wooden swords and spears to start with, going on to blunt metal and then heavier than average ones, so that they'd find the actual swords and spears easier to manage in the arena.

The night before the contests, the manager would lay on a feast for the fighters. In earlier times, the manager or trainer would also have been the owner but then the government took over on the grounds that it was too dangerous to have one man in charge of perhaps 5,000 trained fighting men.

The show started early in the morning but there were purely 'circus' acts to begin with, mostly involving animals doing tricks. After that came the beast fights. Armed men tried to slay the animals without getting crippled or killed themselves. The crowd, however, were impatient for the real 'games' to start, when they could bet on the contestant of their choice and shout or scream their approval as one man after the other was butchered.

There were exhibition contests first, followed by a trumpet fanfare for the parade of the gladiators. They saluted the emperor in his private box and chanted,

Gladiator armour

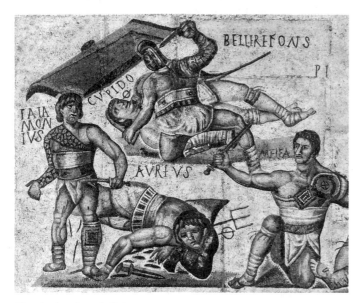
Mosaic showing gladiators

Types of gladiator – the heavy-armed secutor (left) and the retiarius

'Ave imperator! Morituri te salutant!' ('Greetings, emperor. Those who are to die salute you!')

Lots were drawn to see which man should fight which. The opponents could be equally armed and matched but the crowd sometimes liked to see unequal contests. For example, the retiarius had a helmet, dagger and three-pointed fork called a trident, in addition to a large net. The other contestant might be a secutor, with sword, shield, visored helmet, metal leg guards and armour for the sword arm (see page 78).

Each stalked the other, looking for an opening. The retiarius hoped to entangle the secutor in his net. If his throw missed, he could pull back the net with a rope which he kept in his hand. At the same time he must parry sword thrusts with his trident and hope to get in a stab or two with his dagger. Other methods of fighting included chariot battles or contests with lassoes.

The winner received a reward of gold or silver. If the loser was still alive but too hurt or exhausted to go on, he could appeal to the emperor who usually left it to the crowd to decide whether he should be spared or not. A good fighter would be helped out to have his wounds attended, but one who hadn't pleased the crowd was slain where he lay and his corpse dragged out. Fresh sand was sprinkled over the bloodstains and the next fight began.

This sort of thing went on for centuries, the number of public days off gradually growing, so that everyone could revel in the blood letting. At one stage, nearly half the year was 'holiday' for the masses of poor and unemployed. Tens of thousands of men were slaughtered to keep them amused and less likely to rebel. The motto of the government seemed to be 'panem et circenses'. This means 'bread and circuses' and was what the government was ready to supply in the way of food and entertainment for the sake of peace on the streets.

Gradually, however, the spread of Christianity in the empire made these exhibitions less popular and in the reign of Honorius the gladiator fights were banned.

The baiting and killing of animals went on, though, and it is possible that the bull fights which still take place in south west Europe are their descendants.

Bestiarius – animal fighter

The Circus Maximus

Romans had always been keen on horse-racing, the more so when the government banned gambling on everything else. The more poor people there were in the capital, the more betting there was; the downtrodden always seem to feel the need to wager more than the rich, even though they can afford it less.

The first races were not arranged or properly organised. Farmers' sons would gather at a stretch of softish ground along the banks of the streams which drained rainwater from the city's hills.

Two wooden posts were driven in the earth some six hundred yards apart and the horsemen galloped from one to the other, round the top post and back again. Anxious citizens cheered on their favourites and made sure the part-time bookmakers didn't run away with the winning bets.

The Circus Maximus started in this way but although the oldest and biggest, it wasn't the only track in or about the city – and, of course, there were many others throughout the empire.

At the 'Maximus', the two wooden posts were replaced with stone and a ridge of earth was set up between them – so like a backbone that the spectators called it the 'spina'. On this spina were set statues of gods thought to be favourable to racing. The spina was rebuilt in stone and temporary wooden stables erected. These were replaced with stone ones, and a set of starting stalls was erected.

Horse-racing was gradually ousted by chariot-racing – each light, two-wheeled vehicle being drawn nor-mally by four horses, although teams of up to ten animals were not unknown.

Once the spectators had stood or sat on the grassy banks, but another improvement brought rows of seats. The stone ones at track level were reserved for import-ant officials, or even the emperor. Then came the wooden seats and standing room at the back. No one knows for sure but it is thought that there were seats for almost a quarter of a million people.

The fans made their bets and eagerly awaited the appearance of the drivers in their chariots. There were usually four vehicles in each of the twenty-four races, the drivers dressing in tunics coloured white, blue, green or red, depending on the faction, or team, they belonged to.

By this time, the whole complex had been enclosed inside walls like the ones round the Colosseum. This was quite an achievement by the builders, for the walls were about fifty feet high and went right round an area of almost half a mile long and over two hundred yards wide.

Drivers moved their vehicles to the starting stalls, first checking that all was in order. The reins were wound round their waists and tied. In case of an accident, the driver didn't want to be dragged round the arena by stampeding horses, so each one was provided with a razor-sharp knife to cut himself free – that is, supposing he could reach it in time.

Each chariot was led into its place and a rope run across the front of the stalls. The official in charge started the race by dropping a handkerchief: the rope was whipped away and the chariots sped forward.

It was an advantage to get near the stone spine because your turning circle at the far end was shorter than anyone else's. But too near could spell disaster – there were always those who were willing to crowd you into the wall or even to put a hub through your spokes.

Each race consisted of seven laps and the drivers always knew how much more they had to do, because a marshal removed one of seven huge wooden eggs at the end of each lap. Later, the laps were signalled by the turning over of a large bronze dolphin.

Drivers could start on this dangerous profession at the early age of thirteen but few survived to enjoy the huge fortunes that could be won. Although Diocles retired in the year 150 A.D. at the age of forty-two with 3,000 wins behind him, others were not so lucky. Fuscus, Crescens and Mullicius, outstandingly success-ful drivers, were all killed in racing accidents during their early twenties.

Perhaps only the bookmakers were the winners – as in modern times!

Chariot race in the Circus Maximus

The citizen soldier and the legion

If we want to find out about the first Roman army we might do worse than ask one of the city's legionaries.

'What can you tell us about the first legions, soldier?' we ask.

'I'm not really a soldier,' he replies. 'Either that or everyone is. You see, we don't have a standing army. Whenever there's danger, everyone who owns property must take his place with the fighting men.'

'Why property owners?'

'Partly because every soldier (even a part-time one) has to provide all his own equipment – a sword, spear, helmet and any armour he cares to wear. Those of us who are well off can afford to have these things made; poorer people can not. Also, because it's thought that those who own farms or estates have most to lose if our enemies win. Therefore they'll fight harder for the city than anyone else would.

'In any case, we all go home and get on with our work when the trouble's over. They do pay a little for the time we spend in the army but our farms provide us with our real income.

'This worked pretty well when all we had to put up with was an attempt to run off some of our cattle or the

Triarii – the rear line of a Republican legion

Republican legionaries building siege lines – these were used in the siege of Carthage

occasional raid on the city itself. Unfortunately, things didn't stay like that. We Romans decided that the best way to protect our city from such attacks was to beat our enemies, take their city and make it over to our way of thinking.

'The drawback to this was that our frontiers slowly widened as city after city was absorbed within Roman frontiers. As the borders got farther from our homes, service in the army grew longer and longer. It became harder to recruit men.

'One of our kings, Servius Tullius, reorganised the army. He didn't solve the problem of distance from Rome but some things were different. Before his time, the call-up to deal with our foes resulted in a force of three regiments, each of a thousand men and each commanded by its own officer, or tribune.

'Tullius ordered that all free Roman citizens should be divided into five different classes, the richest at the top and the poorest at the bottom. All were still to supply their own kit but the top group naturally supplied most, and it was from the wealthiest section that the cavalry was recruited.

'The outcome of all this was that Rome now had four infantry legions which fought in a kind of phalanx. This was a formation we copied from the Greeks, each man being armed with a long thrusting spear. The legion was arranged in six ranks of five hundred men apiece.

'The next development was a reduction in the number of lance users. These were called "triarii" and were usually elderly men. The middle-aged (say, 30–40 years old) were the "principes" who formed lines in front of the triarii. The "hastati" were the young men who lined up in the very front of the army. Apart from the lancers, every man had javelins for throwing, plus a sword and shield for close fighting.'

We thank our informant and go on our way. We know that long after his time, the phalanx idea was abandoned. Recruiting had been difficult enough when the soldiers had been compelled to walk long distances before fighting: now they often had to sail abroad before taking part in the war and it proved almost impossible to attract enough men to the legions.

Because of this, entry to the army was thrown open to anyone, however humble a citizen he was. Payment was raised and weapons began to be mass-produced. Now, all the soldiers could be armed and armoured alike.

Thousands of poor men (that is to say, the new recruits) far from wanting release, hoped they'd be kept on as long as possible. Their pay, although small, was more than most of them got as civilians.

At last, all the legionaries came from the poorer classes and the foundations of a paid, standing army had been laid.

Uniform, weapons and tactics

There was little uniformity from man to man in the first legions – soldiers could have whatever weapons and armour they liked and could afford. They had thick leather jackets, or perhaps scale or chain armour, if they were rich enough.

Mail was made from sheets of iron and iron wire. Little quarter inch rings were cut or punched from the sheet iron and joined to others by passing an open ring of wire through the ones to be connected and then closing the ring with pliers. Every ring was thus intertwined with its neighbours on each side and at the top and bottom.

Scale armour was a series of smallish iron or bronze plates, each one wired to its neighbours and sewn on to a stout undercoat – rather like the tiles on a roof, or (as the name suggests) the scales on a fish.

Towards the end of the republic, there were changes in the army pattern. We've seen how legionaries were no longer armed with long jabbing spears, nor were they sorted out into battle formations according to age. Every man was paid adequately and given identical armour and weapons.

The most important difference was the introduction of plate armour. It was made from curved plates of wrought iron or steel. The various picccs went round the chest and over the shoulders, being connected with hinges, or fastened together (and on to the body) with leather straps and buckles – perhaps even with simple hooks.

A charge by an Imperial legion

Helmets were hammered from sheet bronze or iron. With their rounded, projecting neck guards, they looked rather like a modern horse rider's cap worn back to front. Nearly all of them had hinged cheek pieces to protect the sides of the face.

The old oval type of shield gave place to the oblong, slightly curved kind, like a section from a cylinder. The new shields were made from three or four layers of thin laths, glued at right angles to each other and forming a kind of plywood. The laths were probably steamed into shape before they were glued. The whole lot was enveloped with leather and a top layer of linen, and the edges were bound with rawhide or hammered bronze. There was a hollow boss in the centre with a wooden handle behind.

A scarf stopped the armour from rubbing the soldier's neck, and a pair of calf-length leather breeches was worn in cold countries.

Each infantryman had a sword hung on a leather strap which went over his left shoulder. A waist belt carried the sheath for his dagger. The two javelins every man had were found to have a drawback when they were first used. They were designed to be thrown, but unfortunately, the enemy could pull them out of the ground or wrench them from his shield and throw them back. Armourers found the answer. It was to make the second, or lower rivet which joined the handle to the head much weaker – perhaps even a wooden, rather than a metal rivet. This meant that the head of the spear struck and then bent over just below the head, rendering it useless.

It is probably true to say that there were more sieges than battles. However, when a battle was inevitable, the Romans drew up their legions in line abreast with cavalry on either side. The idea of upper-class horse soldiers had long been abandoned: the general concerned preferred to raise local auxiliary horsemen whilst on campaign.

Legionaries had once been organised in centuries of 100 men, each under its own centurion. Later it was found more convenient for a centurion to have in his command only 80 men. Six centuries made a cohort and there were always ten cohorts in every infantry legion. The men lined up side by side and about six feet apart. The line behind them covered the gaps thus formed. The legionaries did nothing until a signal from the commander was relayed by trumpet.

Then they threw their javelins, drew their swords and charged. When in contact with the foe, they closed ranks to plug the initial gaps. The front line fought with sword and shield, the fallen being replaced by men from the second line.

Romans didn't have better weapons or superior commanders but they did have discipline. In a scrappy, hand-to-hand engagement, it would have been easy for the faint-hearted to slip away. This was all but impossible in a legion, where years of training kept the men in rows and in contact with the enemy. In any case, the average legionary had the confidence to smash through the opposing lines, knowing that his sides and back were protected.

Roman generals depended on the infantry to cave in the enemy centre, although they did have cohorts of archers and slingers if they were needed.

A legion cohort of the early empire

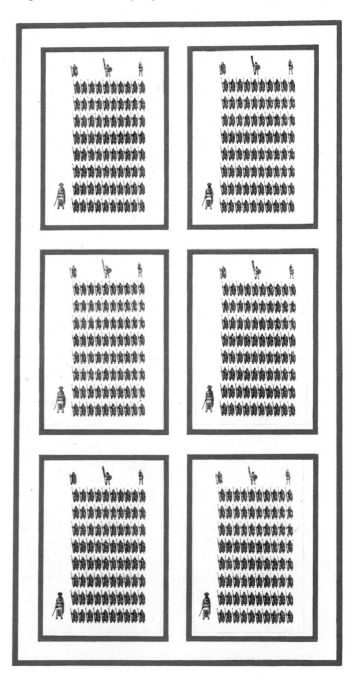

Siege engines

It wasn't only in Rome that sieges were more common than battles. The result was that most of Rome's enemies were familiar with the idea of an army investment of their town and had built stout defensive walls all round it. Army commanders had to know how to reduce an obstinate town or fortress. The emperor Vespasian, when still a legionary commander during the invasion of Britain, stormed and took no fewer than twenty British hill forts in the south of England.

One simple, if boring, way of capturing a town was to surround it and wait for the people inside to starve. Their submission might, of course, take months. Roman officers came early to experience siege methods from the Greeks or from those who had copied them. The ideas had probably originated in the Middle East and they were not improved until the invention of gunpowder in the Middle Ages.

A Roman general, faced with the task of taking a fortress, could order his men to go under, through or over the walls. His choice depended on local circumstances.

Soldiers tunnelled under walls to bring them down, rather than to make a way through for masses of men. Making a hole in the wall could be done with a battering ram. This was often a large tree trunk with one end cased in iron and slung on chains or ropes from

Siege ramp

Large stone-throwing catapult

Vertebrae with embedded
catapult bolt head from
Maiden Castle, Dorset

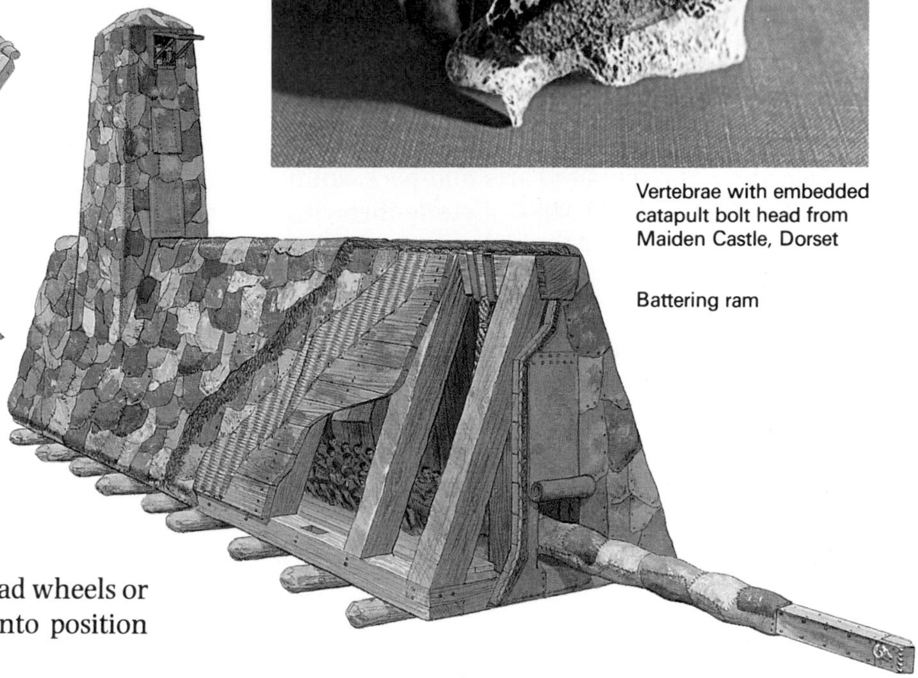

Battering ram

the roof of an open-ended shed. The shed had wheels or rollers and could be moved fairly easily into position against the walls.

It had to be big enough to hold twenty or thirty men whose job was to swing the log against the bricks or stones. To prevent the defenders setting the shed on fire, it was armoured with metal plates or at least covered with rawhides.

Screens of hides also covered wickerwork frames to protect archers as they fired at the enemy. Galleries rather like the battering ram sheds but not quite so stoutly built, were open at both ends and could be put together by the dozen to make a protected corridor for the attackers to use to get to the base of the walls. The corridor could be long enough to start out of range of the fort's weapons, which was quite an advantage. In an emergency, Roman soldiers could cover themselves, top and sides, with their shields to make what they called a 'testudo' (tortoise), if no other protection was available.

Sometimes a Roman commander would order the building of siege towers to overtop the town walls. These were also protected against fire. They could be wheeled up to the walls as well. If necessary, a protective moat or ditch could be filled in to make a causeway for the towers. The infill consisted mostly of logs with clay in between the layers. Sometimes the earthworks thus formed were enormous. During the siege of Jerusalem, legionaries are supposed to have chopped down every tree within a dozen miles of the operation.

In order to clear the enemy away from the walls whilst the rammers battered, the miners dug or the assault troops crossed to the battlements on the drawbridges of the towers, Roman generals called upon the artillery. The picture shows one of the machines that could be pressed into service. Although such 'guns' could be made on the spot, they were far more likely to have been built elsewhere and transported to the required place.

There were stone-throwing machines varying in size from those able to project a stone the size of a tennis ball to large ones capable of shooting a hundred pound boulder almost half a mile.

Other machines were designed to fire small arrows, or heavier ones up to twelve feet long, as far as the large stones could be slung. One dangerous device was to take something that would burn easily, soak it in naphtha or oil, set fire to it and then shoot it into the enemy fortress.

Evidence of the use of catapult arrows was discovered when Maiden Castle was excavated. This was a huge hill fort, taken by the Second (Augustan) legion during the Claudian invasion of Britain which began in 43 A.D. The diggers found a skeleton with a Roman ballista bolt wedged into the vertebrae of a defender's backbone.

They didn't have gunpowder but Roman artillery was nearly as deadly as any gun of the Middle Ages.

Marching camps

The one thing which could slow down the advance of a legion was the mass of carts and pack animals which carried tents, poles, parts of stone-throwing engines, spare armour, weapons and rations. For this reason, officers in charge of moving the legion gradually transferred a great deal of the baggage on to the shoulders of the ordinary legionary.

As well as his armour, spare boots, helmet, sword, dagger and javelins, each man carried a change of clothing, rations for a fortnight (mostly grain of some kind) and a cooking pot. He also had to sling on his back at least two fencing posts, a spade or entrenching tool, a length of rawhide rope, a saw and perhaps a wooden mallet for driving in tent pegs. In total the load weighed almost a hundred pounds, about as much as a sack of coal.

Towards the end of the afternoon an advance party from the marching legion would go forward to seek a camp site for the night. They looked for a flattish area, perhaps on slightly raised ground or a hillside. The site had to have space for a camp some half a mile square;

A centurion laying out a camp with a groma

there had to be fresh water at hand and grass for the various animals. If at all possible, the chosen spot must not have too many rocks, bushes, trees or broken ground which could give cover for an enemy's surprise attack.

When a site had been found, a surveyor, using an

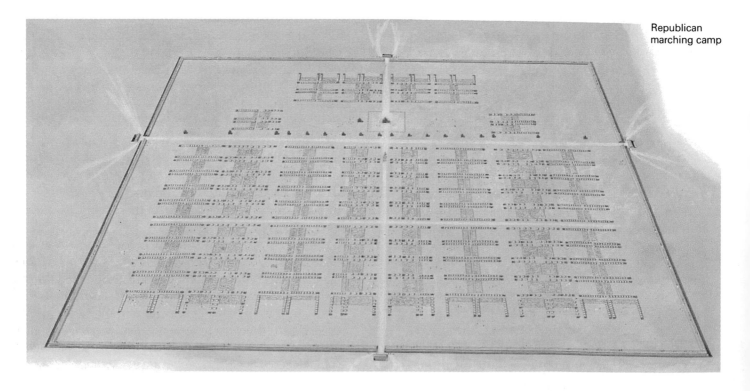

Republican marching camp

Digging trenches for a camp

instrument known as a 'groma', decided where the corners of the camp were to be. These were marked with small coloured flags and the line of the defences scratched through the turf.

Other flags showed the future positions of the officers' and official tents. Another furrow was drawn about two hundred feet inside the defence line. No tent was permitted inside this margin, thus allowing space for the defenders to manoeuvre and keeping tents out of the enemy's artillery range.

When the legion arrived, men knew exactly what their jobs were. Many of them set to work to dig out the ditch. They left a gap in the middle of each side trench for the entrances. A road was laid out between the north and south gateways and another one crossed the camp from east to west. There were no actual gates, the way in being protected by an additional section of trench a little way in front of the opening.

Where the roads met, a space was left for the commander's tent. The space was square, measuring two hundred feet on each side and was called the 'praetorium'. To the left and right of it were two empty areas – one called the 'quaestorium' where some rations were issued and the other a forum for public meetings and parades. Other official tents were nearby.

The ditch diggers threw the excavated dirt just to the inside of the ditch, which was considered adequate if it was seven or eight feet wide and five feet deep. If a serious attack was expected, the ditch was made much deeper and wider. The soldiers all planted their fencing stakes on top of the new bank of dirt and fastened them together with rope.

The men's tents were put up in lines parallel to the defence works. Often, a small unit had its tents and stables erected on a horseshoe plan, with the centurions camped at the two ends of the formation. The tents themselves might be of canvas over wooden poles but were just as likely to be of thinnish leather.

Even when the ditches were finished and every tent erected and anchored with iron tent pegs, the soldiers' work wasn't done. If they weren't unlucky enough to be picked as guards, the men still had to light their fire and prepare an evening meal. Finally, even the special duty men in tents behind the commander's had their porridge and retired for the night. One man from each group had to report to the commander to be given the password for the night.

If the inspecting officer found a sentry asleep at his post, the offender could be beaten to death – punishments were very hard in the Roman army.

In the morning, the commander ordered the trumpeter to sound a call. On hearing it, the working parties pulled up the pegs around the officers' tents and dismantled them. Then the legionary tents were also struck. On a second trumpet call, the tents were loaded on to what carts there were or strapped to the backs of pack animals. A third signal gave the order to assemble in marching order.

Many of these temporary camps became permanent in some frontier towns. The tents were replaced with buildings of brick, stone and timber, and the earth ramp gave way to stone walls and gatehouses.

The officers

The commander of a legion was called a legate. He was actually a politician, usually appointed by the emperor, but needing army experience before he could advance his civilian career. If he was lucky he might be promoted to a provincial governorship after his time with the legions. A man with such a job would have to be foolish or unfortunate not to become rich, so it was worth roughing it in the army for a few years.

As a legate, he wore an elaborately embroidered tunic in scarlet wool, plus a cloak of similar material. His breast plate was made of bronze and hammered out until it resembled the exaggerated muscular development of a professional athlete.

His second-in-command was a senior tribune. In spite of this description, he wasn't necessarily very old. He was also hoping to be a politician – perhaps a senator – and could not expect to succeed without a spell in the army. His uniform was like his legate's but not so elaborate.

In the legion there were another five junior tribunes. These were not normally aristocrats but came from the same social class which provided a good many local government officials. Again, their uniforms were similar to their superiors', if not quite so fancy. On the whole their duties were more concerned with the running of the legion than active service. However, in an emergency, they could (and did) take command of auxiliary or other troops.

Next in seniority came the camp prefect. He had commonly served at least thirty years, much of the time as senior centurion. His main responsibilities were to see that the legion's equipment of all kinds was up to date and in good repair; that every section was at full strength and properly organised; and that training was being done correctly. At a pinch he could command the legion, if he was the only senior officer present but as an ex-centurion, he couldn't normally expect to be promoted any higher.

The average legion consisted of ten cohorts, each of which was divided up into six 'centuries' of 60–80 men apiece. The non-commissioned-officers, or N.C.O.s, were called centurions because they had once been in charge of a hundred men – now they looked after a smaller 'century' each. Centurions had a fancier tunic than a common soldier's over which a leather coat was worn. Body protection was provided by scale armour, guards for the lower legs and a legionary-type helmet, except that the crest went from side to side rather than front to back. Over his coat and armour he wore leather straps on which were his medals and decorations.

It was as well for the ordinary soldier to keep on the right side of his centurion. Men signed on for twenty-five years and were given Roman citizenship as a reward when their time was up. This was a prize worth having for there were all kinds of drawbacks to living in the empire without being a citizen.

Some centurions were hard to the point of brutality, thrashing their recruits unmercifully (as part of their training) with a vine stem, their badge of office.

If you were in the centurion's bad books, you might find yourself constantly punished or even dishonourably discharged with no citizenship. The authorities seemed powerless to deal with the frequently made charge that a centurion could be bribed to overlook a man's offences.

Punishments also included stoppage of pay and leave, or reduction in seniority. For serious crimes like desertion, cowardice, theft or sleeping on guard you could be put to death.

The highest rank an ordinary citizen in the army could rise to (apart from camp prefect) was senior centurion – 'primus pilus', or 'first javelin', as he was called.

Other minor officers below centurion rank were known as 'principales'. The 'optio' was the centurion's second-in-command. There were also the standard bearers – either the 'aquilifer', who carried the legion's eagle, their main badge, or the 'signifer' who looked after the century's own standard. At the bottom of the command structure and only just above the lowest rank of legionary were vets, doctors, musicians and orderlies.

Officers in a Roman legion

Left to right:
Standard bearer for century (signifer)
Aquilifer with eagle – the legion standard (behind)
Centurion
Centurion horn blower (cornicen)
Trumpeter (tubicen)
Commander-in-Chief (Provincial Governor – Legate) (seated)
Legion Commander (Legatus legionis) (behind)

Roman roads

Today we take roads so much for granted, we scarcely think about them. If we want to travel about, no matter by what method – car, bicycle, bus or whatever – we don't have to think whether we can get to our destination easily. Of course we can: we can look at a map and pick our route.

Now let's try to imagine that there are no roads at all. Immediately, we realise that we can't travel by car or bicycle but only on foot. If you've ever tried running up the side of a sand dune or through heavy mud, you'll know what a tremendous difference is made by the lack of a hard surface under your feet.

When you went on a bus, coach or car, did you ever cross a river? Would you even have noticed? If you were walking, you'd quickly learn to look for the easiest way to cross even a small stream.

If it's a task for you to make your way through a forest, over a marsh, up hills and across other difficult countryside, just think how much harder it would be to try moving something heavy over the same route.

It's probably true to say that if the Romans had been unable to bring in food and building materials on specially made roads, Rome could never have grown to the size it did.

In the early days, only faint tracks ran from the infant city to nearby towns and villages. Then in 312 B.C., the censor, Appius Claudius had a proper road built from Rome to Capua. Other roads followed as further conquests were made, for example, the Via Flaminia which ran north to Rimini.

Roads were also a military necessity. If there was trouble, troops could get to where it was without delay.

The main roads of the Roman Empire

As we've seen, marching can be much faster on a good hard surface. Every new city conquered had a road built to it – 'All roads led to Rome'.

In the reign of Augustus, a golden milestone was set up in the 'Forum Romanum', the market place and main square of Rome. The distances of all other towns were measured from it. There were occasional milestones on the new roads. These were cylinders of stone standing five feet or so above ground level, showing how far the next town was and which emperor had ordered the stone to be erected.

Roman roads were as straight as possible. If a high cliff or a mountain was in the way, the road was angled round it – but if the obstacle was merely a hill or a fen, then the engineers went straight ahead.

They would pick a spot on the horizon – a tree, perhaps. If there was nothing to see, an advance party would be sent ahead to light a fire in the distance to provide a point towards which the road must be driven.

The roads were made both by civilians and soldiers but perhaps the last named did most. A lot of the heavy labouring was done by slaves.

The outline of the proposed road was marked with poles and the road bed cleared by pick and shovel. Then various layers of mortar, stone, clay and gravel were laid bringing the surface up to about three feet above the old level. In places where traffic was heavy – for example, near large towns – the engineers ordered flat slabs of stone to be laid on the top. Most roads had a slight hump or camber in the centre of the highway. When it rained, the water trickled into the ditches running on each side. The drainage ditches were about twelve or fifteen feet apart, although some roads were wider and others narrower. The rule seemed to be that there must be room on a main highway for two legions to pass one another.

Along the roads travelled not only soldiers but ordinary people and goods on all kinds of journeys. If the emperor wanted a message sent, his messenger rode a horse: if he rode furiously, changing mounts every few miles, he might cover two hundred miles in a day. He couldn't keep it up however, but if he had been able to, it would still have taken nearly a fortnight to get from one end of the empire to the other.

This may sound slow going but after the empire collapsed, it was to be thirteen centuries before man could move as fast again.

Roman roads covered most of Europe and the lands around the Mediterranean in a complete network. There were enough main roads to have gone round the Equator twice and sufficient roads of all kinds to circle the world ten times!

Troops of Emperor Augustus building the road over the Great St Bernard Pass

Barbarians settle inside the frontiers

The groups of men in front of us are not speaking Latin, although we are several miles inside the frontiers of the Roman Empire. Not only that – they are armed to the teeth with spears and swords slung from leather baldrics or shoulder straps, and they are wearing conical helmets and sheepskin jackets.

We know that for centuries, the Roman army has recruited men from anywhere in Italy and finally from anywhere in the empire: but these men aren't dressed in uniform and they don't look like legionaries from a standard unit. Perhaps we could ask where they are from? We beckon and one of them strolls over, his hand resting lightly on the hilt of his sword.

'We'd like to know who you are and what you are doing inside the empire. You obviously aren't in a legion – or are you?'

The man laughs. 'No, we're not Roman soldiers. We come from the north and east. The Roman authorities let us come in and settle this side of the frontier.'

'Why should they do that?'

'I can't really say for certain. All I can do is guess. You see, we were living peacefully – well, as peacefully as anyone can in these troubled times. Then, only a year or so ago, our neighbours began to raid more often than they'd ever done before.

'Once upon a time, all they wanted to do was to make a surprise attack and run off some of our horses and cows. Then they began to press us back – not just a cattle raid, you understand, more like an all-out attempt to push us out of our territory.

'We captured some of their horsemen and demanded to know why we were being thrust back. They told us that they had already lost almost half their own north-eastern lands to a people which had come from the far

The Rhine/Danube frontier of the Roman Empire, showing incursions by barbarian tribes

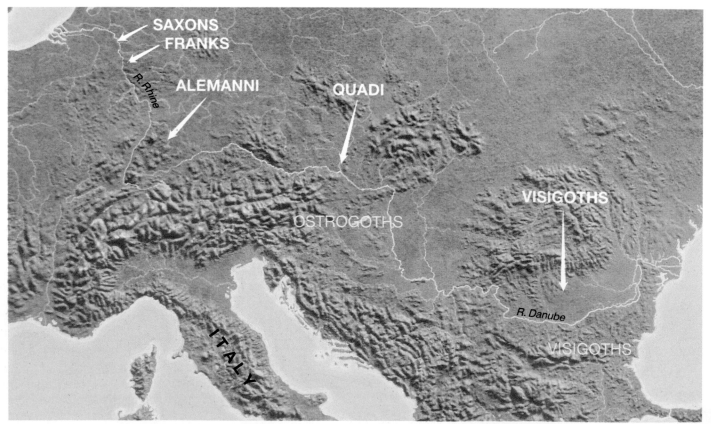

Gothic archer and horseman

Hunnish horseman

east. These people all rode on horses and used short but very powerful bows. Captured bows revealed that they were made of thin strips of bone, glued and pinned together. One of our prisoners swore that he had seen one of the arrows go straight through a shield.'

'Do you know what these invaders were called?'

'They called themselves Huns. They were quite strange looking. They had brownish-yellow skins, almond shaped eyes and high cheek bones. They also carried wickedly sharp sabres. A new feature was their use of murderous lances – from horseback, naturally – but they couldn't have been effective without stirrups. You look surprised?'

We nod, so he continues, 'If you hit something with a spear when you're riding, you're more likely to come out of the saddle without stirrups.

'Anyway, these Huns were driving our neighbours back; our neighbours attacked us, so we asked the nearest Roman governor to let us settle on the south side of the Danube. The land we wanted was only thinly populated so we thought we had a good chance.

'To start with, the Romans wouldn't listen – until we made them an offer they could not refuse. We offered to defend the banks of the river ourselves and keep out any other "barbarians", as the Romans call us.

'It was obvious that the governor was at least interested. He told us to go away while he consulted with his colleagues. So we waited. We found out later that other tribes from beyond the empire had put forward the same kind of proposal and in some cases the Romans had accepted.

'Finally the governor gave us his permission – but there were conditions. Some of us had to join their regular army and the rest of us were divided into smaller groups and kept apart from each other.

'I don't think the Romans are too happy about the arrangement, though, and I know that in some instances, they preferred to pay us savages just to go away. But there certainly are similar tribal groups settled on the "wrong" side of the Rhine-Danube frontier in several areas. They do say that there are more of us inside the empire in other areas. I can't swear to that but I must say that if I were a Roman, I wouldn't feel too secure relying on what to them must seem like enemies to guard their precious empire.'

Some provinces are abandoned

The first reason why Rome had expanded was that other cities had attacked her. The best defence is attack and early Romans might have said, 'I'm not a bully, nor am I greedy: all I want is the land that touches mine!' This motto might have been the reason why Rome went on growing.

However, during the reign of Trajan, the empire had become as vast as it was ever to be. Hadrian, the emperor who followed Trajan, thought that it was already too big and voluntarily gave up a large part of the eastern provinces. This meant that the expansion had at last come to a halt and that from then on a slow shrinking had begun.

The barbarians around the frontiers pressed forward more and more often. The defence of Rome wasn't helped by the fact that there was no satisfactory way of getting rid of a bad emperor, short of killing him. Soldiers, rather than the senate, were the emperor makers and if they weren't rewarded by the man they had chosen, they murdered him and looked around for someone else.

In our own twentieth century, there have been no more than four British rulers during the last seventy years. During the same length of time at the beginning of the third century, Rome had twenty-three, only one of whom died a natural death.

Among the barbarians now battering at the imperial frontiers was a group of Germanic tribes, including the Franks, the Alemanni and the Goths. These were crossing northern and north-western Roman lines, whilst other outlanders were doing the same in the east.

To make matters worse, the empire was rapidly becoming harder to govern, as law and order collapsed and street riots became more common. At times there was even civil war within the empire. To deal with this, legions were withdrawn from the fighting and sent to Rome to keep the peace. As a result, the defences were considerably weakened and in Dacia, to the north of the river Danube, there were hardly any legionaries left. The Goths poured across the abandoned lines in their thousands.

One Roman emperor fell in battle, one was captured

Aurelian

and many others assassinated. They came and went swiftly as various army commanders vied with each other to take the throne. At one stage, no less than nineteen senior officers strove to be the next emperor.

Among the crowd of would-be rulers, a few names are outstanding in the fight against the outlanders. The emperor Aurelian, for example, actually defeated a Gothic army, although he did allow them to settle on the Roman side of the boundaries, perhaps hoping that they would keep out any further adventurers. He beat back another invasion of the Alemanni, put down rebellions in the east and generally restored law and order. It is a pity there were few emperors such as Aurelian. But in spite of his good record at protecting the empire, he too was murdered, as were both of the next two emperors.

No wonder provinces were abandoned under the pressure of waves of advancing barbarians – there was hardly a Roman ruler who lived long enough to think up a reasonable defensive plan and put it into operation.

Emperor Aurelian built walls around Rome in about 275 A.D. They had square towers every 100 feet and were 12 miles long.

Constantine, his town and the division of the empire

When Diocletian ruled Rome at the end of the third century, Rome was already more than a thousand years old and much too large to govern properly. The emperor therefore decided that everything would be easier if the whole area was split in two.

He promoted one of his men to be co-ruler. This was Maximian who would rule the western half from Milan, while the emperor would run the eastern part from what is now Turkey. Many of the citizens of Rome must have found the new arrangements very strange. Let's go back to the fourth century and ask someone about the changes.

'Excuse me, are you a citizen of Rome?'

'Of course I am. Everyone knows me: my name's Marcellus.'

'But do you live in Rome itself?'

'Yes, I do have a home there but I've two or three other homes as well. I began work as a sailor, rose to be a ship's captain, and after a while bought a ship of my own. Now I've got a fleet of 'em. We carry some passengers but mostly we take cargoes such as timber, wool, wheat, iron ore, pottery and even wild animals for gladiators to fight.'

'What happened when the empire split?'

'A blow to the pride of those who live in Rome, perhaps, but no great differences in day-to-day matters.'

'What exactly were the alterations?'

'I suppose you already know that Diocletian and Maximian shared the ruling between them? Yes? Well, each man was called "Augustus" – after our first emperor, of course – and each had an assistant known as a "Caesar".

'Diocletian and Maximian did something rather surprising a few years later. The year 305 it was.'

'What did they do?'

'Why, they resigned. Sensible really. No point in waiting about to be stabbed to death, eh? Yes, they retired and Diocletian went to live in what you call Yugoslavia. He built himself a fine palace at a place called Split and became interested in growing cabbages, of all things.'

Constantine defeats Maxentius, son of Maximian, at the Battle of Milvian Bridge

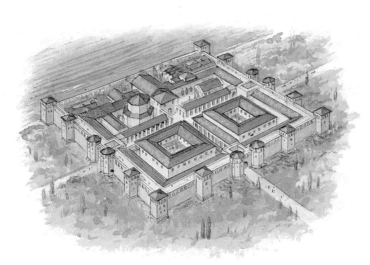

Diocletian's palace at Spoleto (now Split) in Yugoslavia

'Who took over from them?'

'At least half a dozen men struggled for power. One of them, a man named Constantine, won in the end. He was put forward by his troops – "proclaimed" is the word they use. It happened at York in Britain, as a matter of fact. He was the one who is said to have seen the vision of a cross in the sky during one of his battles. I don't know if that's true but from then on, people were allowed to be Christians openly.'

'Didn't Constantine make a new capital city?'

'He certainly did. Maybe he wanted to get away from those men in Rome who still thought they ought to have some say in the government. The emperor didn't want his word to be questioned so he made the new chief city at a place just about as far from Rome as he could get. Byzantium was his choice. It's an old Greek city standing on the narrow strait which divides the Mediterranean from the Black Sea. It must have seemed a miraculous place to Constantine: it was already a Romanised town with the usual temples, market place, baths, theatre and so on – but what astounded him was the fact that it even had seven hills, just like Rome itself.

'Constantine began enthusiastically in 326 to change it into a great city. It was called "Constantinople" from then on. I suppose that must be a sort of Greek word meaning "Constantine's town".'

'I think we call it "Istanbul" nowadays', we say.

'A lot of people called it "New Rome". It was half in Europe and half in Asia, with a good anchorage for ships. Not just naval ships, you understand, but merchant vessels like mine.

'I suppose the biggest difference it made to me was the easing of trade between east and west. And, of course, more profit,' he smiles. 'But whether it'll be a good thing in the long run, I can't say. It's still rather odd to feel that Rome is no longer the most important town in our empire, and that's a fact.'

Constantinople from the air

Alaric and the sack of Rome

We've seen that some emperors sought to preserve the frontiers by allowing in some of the barbarians and bribing them to keep out any others. The leader of such a group was Alaric, the chief of an east European tribe of Visigoths (western Goths). The Visigoths had probably been forced westward and into what we now call Bulgaria by another barbarian tribe called Huns. Alaric was born about the year 376 into an important family then living on an island at the mouth of the river Danube.

As a young man he was appointed general of a group of irregular troops who helped the emperor Theodosius to crush a rebellion. He was hoping to be given a proper commission in the regular Roman army as a reward but the old emperor died and the new one, Honorius, did just the opposite. Not only did he refuse any reward, he also cut off the 'presents', or bribes, which had bought the service of the Visigoths.

Alaric's men then proclaimed Alaric King of the Visigoths and together they swore to carve out a new

kingdom for themselves. Alaric led them towards Constantinople only to find the place was far too well defended to fall to their attacks. They headed for Greece, capturing and looting one city after another and selling the inhabitants into slavery. A Roman regular army cut off their retreat and they had great difficulty in fighting their way free.

This was almost the last victory of proper Roman legionaries fighting on foot with javelin, sword and shield. In the old days, such men were very nearly unbeatable. At the time of Alaric however, Rome now depended on legionaries who – far from being citizens of Rome or dwellers in Italy – were not even born within the boundaries of the empire. They were, in fact, largely

Alaric and his Goths sack Rome. The Basilica Julia (right) and the Temple of Divine Julius (left) were burnt.

recruited from the very tribes of barbarians which threatened Rome's existence.

In the old days, Romans had fought for their own families and homes, but now the empire was being defended by men bribed to fight for others. Woe betide Rome if the bribes were not handed over.

At one stage, Alaric and his men seemed so dangerous to Romans that they were bribed to protect the lands which lay between the western and eastern empires, now no longer on friendly terms with each other. Alaric played off one Augustus against the other, even managing to get official government factories to provide weapons and armour for his men.

Still smarting from the refusal of the authorities to give him respect in the form of an army commission, Alaric decided in 400 to attack Italy itself. He was defeated after a couple of years but the strange thing was that the army which drove him away was also led by a barbarian. His raids had the effect of causing the emperor to move his capital and (rather more important as far as Britain is concerned) to bring back a legion from Britain in order to defend the heart of the empire. In addition, the tying up of Roman armies in this struggle had allowed other tribesmen to sweep across Europe, occupying both France and Spain, which then passed out of Roman control for good.

Twice more Alaric attacked the city of Rome and on his third attempt, his Visigoth warriors broke through the gates and into the streets of the old capital. As a result of the attack, Romans in North Africa cut off supplies of corn to the city. Alaric thought that the only way to deal with this was to attack the North African provinces in order to get the grain ships sailing again. During the voyage across the Mediterranean, a fierce storm arose, most of the ships were wrecked and the Visigoth army they were carrying was almost wiped out.

Alaric himself died shortly afterwards – some say of a fever – and his body, together with his weapons and armour, was buried in the bed of a river which had been temporarily diverted. When the funeral was over the river was allowed back in its old course and the slaves who had done the work were slaughtered to stop them telling anyone where the grave was.

All this should have been good news for the Romans, but during Alaric's life, Rome had been attacked repeatedly, the central part of the empire lost, France and Spain overrun by other barbarian peoples and the eastern and western halves of the old empire reduced almost to a state of war with each other.

It wasn't quite the end of the Roman Empire, but it was the beginning of the end.

More barbarian attacks

There was more trouble after the death of Alaric. Rome tried hard to stem the inward flow of wild tribesmen but in 429, Vandals captured Carthage and the rest of the North African Roman provinces. In 410, the Romanised inhabitants of Britain had appealed to the Roman emperor for help against the hordes of barbarians that were attacking Britain. The emperor's message was to the effect that they must look to their own defence as Rome was unable to send troops to fight off the attacks. In fact no troops were ever sent again.

Some towns and cities in Britain were still standing after the invading Angles and Saxons moved in but the outlanders had little knowledge of building or civil engineering and so allowed the places where Romans had once lived to fall slowly to pieces. Even if the newcomers had wanted to use villas and temples they

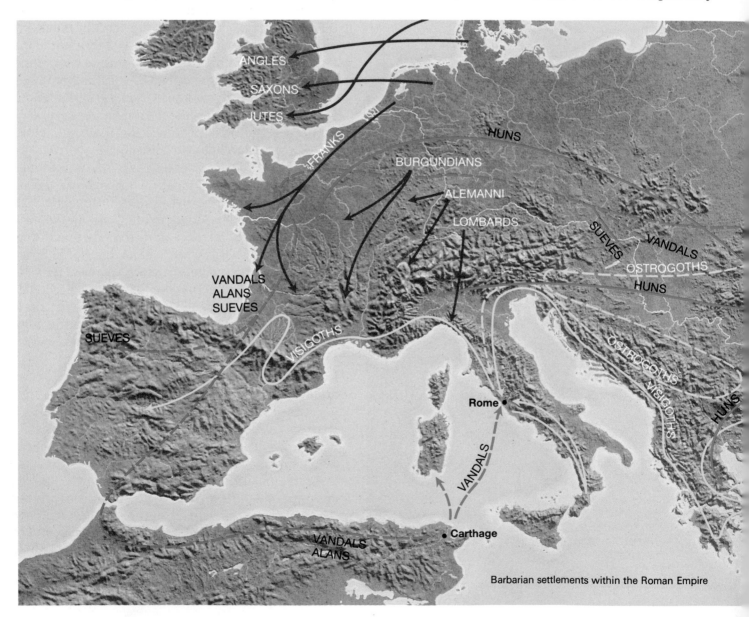

Barbarian settlements within the Roman Empire

had no way of repairing and protecting them.

The same thing applied to roads. Roman engineers had left Britain with a fine network of highways and lesser roads, which the incoming Angles and Saxons were happy to use. This was fine provided that nothing went wrong – but if a flood washed away a surface or a tree fell across the road, the local people who lived nearby felt no need to go and make good the damage. Travellers coming to the obstacle did nothing either, they merely walked round it.

Such authorities that were still in existence in Rome had more pressing problems than what to do about the outlying province of Britain. In March 451, more than 70,000 of a new tribe of barbarians crossed the Rhine into France capturing cities such as Paris, Metz and Orleans – burning, killing and pillaging as they went. These were the Huns. The only way such outlanders could be stopped was by another army of outlanders who might still be faithful to Rome.

The leader of the Huns was a man feared throughout the civilised world. His name was Attila and he was described by a writer of the time as being 'short and thick, swarthy with grey hair, deep-set eyes and an upturned nose. He walked with a proud swagger as though he was the greatest lord who had ever lived.'

This time the Huns were turned back but they returned in the spring of the following year, laying waste to much of northern Italy. They captured many cities including Padua, which they burned. However, the Pope came out to meet Attila at the head of his army of hard-bitten horsemen. No one knows what was said when they met, but the Pope apparently talked the Hun leader out of attacking Rome. Perhaps Pope Leo reminded Attila what had happened to Alaric, the last barbarian to do so. It may be that Attila was superstitious and didn't want to die immediately afterwards, as Alaric had done. It's more likely that he feared that famine and plague then affecting Italy would weaken his army until they were unable to fight their way back over the Alps. At all events, he turned his back on Rome and headed north once more.

Death of Attila

The next year, Attila got married. He ate and drank so hugely at the wedding feast that he burst a blood vessel and died. The menace of the Huns faded away, never to rise again.

But Rome's end was very near. Another band of Vandals sacked Rome in 455. The looting lasted a fortnight. Barbarians were now permanently in charge of Rome and her affairs.

In 476, the very last Roman emperor was deposed. Oddly enough, his name was Romulus, the same as the first king and founder of Rome over 1200 years before.

By the end of the 400s, both Italy and the rest of the western empire had been shared out between the various barbarian tribes – the Suevi and Visigoths in Portugal and Spain, the Franks and Burgundi in France, the Saxons and Angles in England, the Ostrogoths (eastern Goths) in Italy and the Vandals in North Africa. The *western* empire was at an end.

The Byzantine Empire lasts another thousand years

To get some idea of what happened to the rest of the Roman world, we'll ask a historian to give us some facts.

'The western empire was all but dead,' he says, 'overrun and occupied by tens of thousands of outsiders. The eastern part was luckier. There, the emperor

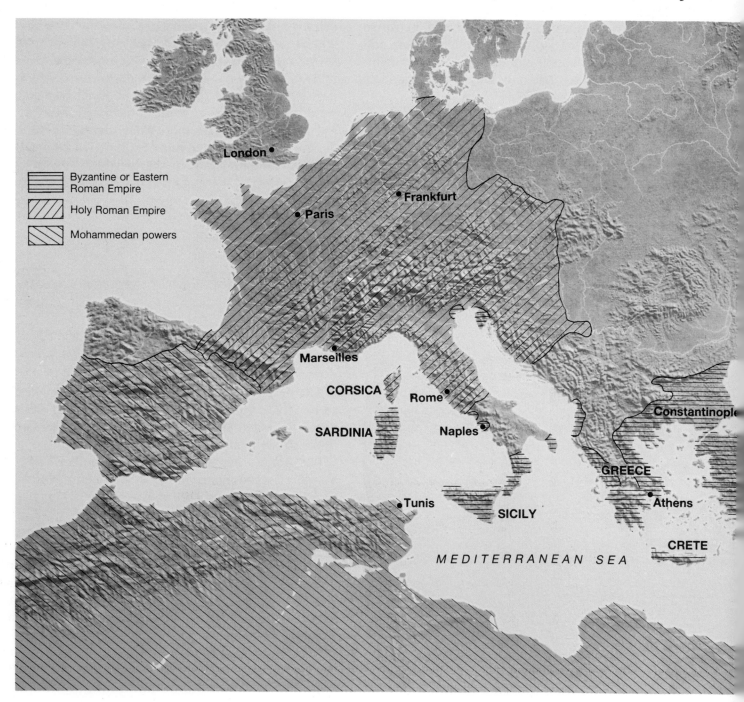

was able either to beat off really serious attacks or to absorb some of the invaders into his territory. He began to build up the army and then finally abandoned the idea that only foot soldiers could be effective.

'Instead, he had his men trained as horse soldiers. They rode into battle in mail coats and were armed with spear and sword. Cavalry of this kind found it easier to deal with the incoming waves of enemy tribesmen, many of whom were also mounted on horseback.

'So well did the eastern empire do at first, that the emperor Justinian in the sixth century was actually

Europe at the death of Charlemagne

CYPRUS

Jerusalem

Cairo

able to recapture some parts of the west that had been lost.'

'The Roman Empire got bigger again?' we ask.

'I'm afraid it was only temporary,' replies the historian. 'The reconquered Roman lands were taken back by the barbarians only three or four years after Justinian's death. Even the eastern empire itself was slowly crumbling away as a result of more barbarian assaults.

'Mohammedans, intent on spreading their religion, overran lands to the east of the Mediterranean such as what are now Egypt, Israel and Syria. Then in the seventh and eighth centuries, they swept through the old Roman provinces of North Africa and across the sea at Gibraltar. They occupied Spain where they were to stay for more than another seven hundred years. They even tried to conquer France but were finally driven out by Charles Martel.'

'Who was he?'

'He was the King of the Franks, the barbarian tribe who then lived in France and from whom the country gets its name. You may be interested to know that his grandson was Charles the Great, or Charlemagne, as we usually call him. Charlemagne ruled almost all of mainland north western Europe and was pleased to call his vast lands "The Holy Roman Empire".'

'But it wasn't – is that what you're saying?'

'No, it wasn't. It was neither an empire nor Roman. I suppose the word "holy" might be excused, since many of the tribesmen that Charlemagne ruled had become Christians.

'It's strange really to think that when they overran what we now know as Germany and France, they probably spoke a language rather like modern German. Those in the German part kept and developed this tongue whilst those in the French part began to speak a common form of Latin which slowly changed through the centuries into French.

'Another group of Mohammedans called Ottomans, or Ottoman Turks, gradually pushed westward into Asia Minor and the Holy Land. Several crusades, or wars of the cross, were begun by Christians of the west to try and drive them out of the Bible lands but all failed in the end.

'Finally in 1453, the Ottoman army, led by by Sultan Mohammed, besieged and bombarded Constantinople until it fell.

'With its capture the last traces of Roman rule in Europe disappeared. The so-called "Holy Roman Empire" of Charlemagne went on – at least in name, if in nothing else – until 1806 when Napoleon did away with the title. Thus was the very last link broken.'

109

Postscript

The legacy of Rome

When we talk about a 'legacy', we are normally referring to some money or property that has come to us in a relative's will. We are usually pleased to have whatever the will maker has decided to leave us. In the case of Rome, it isn't money or property but rather a legacy of ideas and ways of thinking and behaving.

What are the things that Rome has left us? It would be as well to list them (although not in order of importance) as follows:

1 *Law* In the early days, this was mostly the result of traditional public wisdom. Laws dealt with civil matters such as land tenure, trespass, inheritance, contracts, etc. Only later did republican magistrates publish the rules – on bronze plates known as the 'Twelve Tables'. During the time of the emperors, criminal activities also came to be recorded with proper punishments. Justinian collected the laws, sorted them out, discarded a number which were out of date and published the rest. Roman law was taken to every part of the empire and forms the basis of the legal systems of many modern countries, including those nations whose existence wasn't even suspected in Roman days – Australia, for instance, and the United States.

2 *Cleanliness and water supplies* Once the empire was overwhelmed, these things which had been taken for granted were swept away, not to return, even to the developed countries of the world, until the last century.

The White House, Washington D.C.

3 *Roads* The legions constructed thousands of miles of roads all over the empire. In many cases they are still being used – either as the base of modern highways, or at least, as 'easiest route' markers for roads and perhaps even railways.

4 *Towns* Many towns began as army camps or as important sites, perhaps at a crossroads or a ford. Rome itself 'just grew', but later 'planned' towns often follow a more geometric square pattern. Some north American settlements have copied the idea in modern times. Many cities have been provided with the equivalent of the Roman forum, i.e. an open space surrounded by imposing buildings, such as Trafalgar Square in London or the Place de la Concorde in Paris.

5 *Buildings* Techniques of various kinds were either invented or at least developed by the Romans, for instance, the self-supporting arch of bricks or stones. Multiple arches made a barrel vault whilst a groined vault was formed from two barrel vaults crossing at right angles. A simple arch turned through 180° produced a dome. Some huge arches were constructed to make bridges and aqueducts.

Other advances include the use of structural

Trafalgar Square, London. The National Gallery is on the left.

concrete, stonefaced walls banded with thin red bricks and packed with rubble.

A lot of eighteenth, nineteenth and twentieth-century buildings are copies of the Roman style – for example, the National Gallery and the British Museum in London, the Arc de Triomphe in Paris and the White House, Washington.

6 *Government* Government systems at national and local level have been copied in numerous places, as has the idea of a well-trained civil service. Some of the original Roman civil servants almost certainly used a form of Latin shorthand.

7 *Calendar* Our arrangements of months and weeks, including the names of the months (and Saturday) come from Roman sources.

8 *Sport* The Romans invented the sports arena with tickets and numbered seats: the word 'sport' itself is from Latin.

9 *Weights and measures* Some British imperial weights and measures were based on Roman ones. The idea of standard lengths, sets of containers and weights was a Roman one.

10 *Miscellaneous* Many other things were first tried out or adopted on the banks of the Tiber – for example, a postal delivery service, hospitals, our own alphabet, central heating, glass windows, blocks of flats, the organisation of the army, banks, etc.

11 *Latin* The language survived the barbarian invasions because it was used in the church and then at universities or schools and in law courts. The local Latin slang developed into what we call the 'Romance' languages – e.g. French, Spanish, Italian, Romanian and Portuguese. An interesting illustration of the way a legionary's slang became the basis of a modern language is provided by the everyday word 'testa'. It meant (literally) a 'pot' but was used to mean 'head', just as we might say 'napper' or 'bonce'. It was used instead of the proper word 'caput'. Then 'testa' changed into 'tête', which is modern French for 'head'.

We still use some expressions which come directly from Latin such as 'cave' ('beware'), 'tempus fugit' ('time flies'), 'et cetera' ('and so on') – and a good half of English words are descended, directly or indirectly from the tongue the Romans spoke.

12 *Numbers* Roman numerals are used in books for chapter headings and volume numbers, on some clocks and occasionally at the end of television programmes.

13 *Remains* There are still Roman ruins scattered the length and breadth of the old empire. Rich young men in the last couple of centuries finished their education by touring to see what the Greeks and Romans had left behind them. European rulers of all periods thought it was quite the thing to have a portrait or a statue made showing them in classical dress (tunic and toga, or Roman officer's uniform), rather than their normal attire.

It's very true that our world would have been a totally different place if the Romans had never existed.

Index

Aeneas 4, 5, 11, 24, 25
Agrippina 49
Alaric 104–5
Alba Longa 5, 6, 11, 12
Alemanni 98, 100
alphabet 22
Antoninus Pius, Emperor 53, 71
Appius Claudius 32
Attila 107
Augustus Caesar, Emperor 48, 49, 71, 97
Aurelian, Emperor 100, 101

barbarians 98–9, 100, 104, 106–7
Brutus 46
Byzantine Empire 108–9
Byzantium 103

Caligula, Emperor 49
Caracalla, Emperor, baths 35
Carthage 11, 24, 25, 26
    at war with Rome 27–9
Cassius 46
Celts 20, 21
chariot racing 12, 85
Charlemagne 109
Charles Martel 109
Christianity 74–5, 103
Christians 42, 51, 74–5
Claudius, Emperor 16, 20, 49, 71
Cleopatra 47
Commodus, Emperor 53
Constantine, Emperor 103
Constantinople 103
crusades 109

Dido 11, 24
Diocletan, Emperor 102, 103
Domitian, Emperor 51, 71

Etruscan culture 16–17, 45
Etruscan people 12, 16–17, 18–19, 21, 45

Franks 98, 100, 109

Galba, Emperor 49, 50
gambling 68–9, 84, 85
Gauls see Celts
gladiators 79, 82–3
Goths 100
Greek culture 16, 22, 23
Greek influence 36, 38, 65, 87

Hadrian, Emperor 52, 53, 71, 100
Hannibal 28–9
Herculaneum 54
Horatius 18–19
Huns 104

Julius Caesar 46, 47, 71
Justinian, Emperor 109
Lars Porsena 18–19
Latin language 111
Lavinium 11
Lepidus 47

Marcus Aurelius, Emperor 53
Mark Antony 47
Maximian 102

Nero, Emperor 42–3, 49, 50, 51, 75, 80
Nerva, Emperor 51

Octavian 47, 48 see also
    Augustus Caesar
Ottomans 109

Pertinax, Emperor 71
Phoenicians 24, 25, 27
Pompeii 36, 38, 54–5, 64
Pompey 46
Pope Leo 107
Punic Wars 28–9, 73

Q. Fabius Pictor, historian 11
Quintus Fabius Maximus 29

Remus 7–9, 10, 11, 25, 73
Rhea Silvia 6, 7, 8
Roman:
    army 86–97
    baths 34–5
    buildings 22, 36–8, 43, 80–81
    calendar 12, 62–3, 111
    culture 64–5
    entertainment 68–9, 78–81, 82–5
    law 22, 71, 110
    life 35, 40–1, 56–61, 66, 68–9
    religion 12, 36, 72–3, 75
    see Christianity
    roads 96–7
    water supply 32–3, 39
    writers 65

Rome:
    city 30–43

    sacked 105, 107
government and society:
    Emperors 42, 48–53
    Empire 50–3; breaking up 100,
        102, 104–7, 108–9
    equites 66
    Kings 12, 45, 70
    patricians 46, 70
    plebeians 46, 70
    Republic 46–7, 70, 71
    senate 48, 70, 71
    senators 46, 66
    Triumvirate 47

Romulus 7–9, 10, 11, 12, 14, 24, 73

Sabine tribe 12, 14, 15
Sabine women 15, 73
Saxons 98, 106
Scipio 29
Sejanus 49
Servius Tullius, King 12, 45, 87
Severus, Emperor 71
slaves 66, 76–7, 97
Spartacus 77
Sulla, 46

Tarquinius Superbus, King 12, 18, 45, 70
Tiberius, Emperor 49
Titus, Emperor 50, 51
Trajan, Emperor 51, 52, 53, 100
Trajan's column 51, 65
Troy 4, 11

Vandals 106, 107
Vespasian, Emperor 43, 50, 51, 80
Vesuvius 54–5
Virgil, Latin poet 4, 5, 24, 25, 65
Visigoths 104